THE RAILWAYS OF BRADFORD AND LEEDS

Their History and Development

THE RAILWAYS OF BRADFORD AND LEEDS

Their History and Development

PETER WALLER

PEN & SWORD TRANSPORT

AN IMPRINT OF PEN & SWORD BOOKS LTD.
YORKSHIRE – PHILADELPHIA

The Railways of Bradford and Leeds: Their History and Development

First published in Great Britain in 2023 by
Pen and Sword Transport
An imprint of
Pen & Sword Books Ltd.
Yorkshire - Philadelphia

Copyright © Peter Waller, 2023

ISBN 978 1 52677 342 5

The right of Peter Waller to be identified as Author of this work has been asserted by him in accordance with the Copyright, Designs and Patents Act 1988.

A CIP catalogue record for this book is available from the British Library.

All rights reserved. No part of this book may be reproduced or transmitted in any form or by any means, electronic or mechanical including photocopying, recording or by any information storage and retrieval system, without permission from the Publisher in writing.

Typeset in 11/13 Palatino by SJmagic DESIGN SERVICES, India.

Printed and bound by Printworks Global Ltd, London/Hong Kong.

Pen & Sword Books Ltd incorporates the Imprints of Pen & Sword Books Archaeology, Atlas, Aviation, Battleground, Discovery, Family History, History, Maritime, Military, Naval, Politics, Railways, Select, Transport, True Crime, Fiction, Frontline Books, Leo Cooper, Praetorian Press, Seaforth Publishing, Wharncliffe and White Owl.

For a complete list of Pen & Sword titles please contact

PEN & SWORD BOOKS LIMITED
George House, Units 12 & 13, Beevor Street, Off Pontefract Road,
Barnsley, South Yorkshire, S71 1HN, England
E-mail: enquiries@pen-and-sword.co.uk
Website: www.pen-and-sword.co.uk

or

PEN AND SWORD BOOKS
1950 Lawrence Rd, Havertown, PA 19083, USA
E-mail: Uspen-and-sword@casematepublishers.com
Website: www.penandswordbooks.com

CONTENTS

Abbreviations 6

Acknowledgements and Author's Note 7

Introduction 8

Bradford area 24

The Midland Lines from Leeds 54

Halifax and the Calder Valley 85

The Heavy Woollen District 100

Leeds Area 111

The North Eastern Lines from Leeds 141

Leeds South and East 154

Engine Sheds 165

Preservation 175

Industry 181

Appendix 188

Bibliography 206

ABBREVIATIONS

BR	British Rail(ways)
DB	Deutsche Bahn (German State Railways)
DMU	Diesel Multiple-unit
EMU	Electric Multiple-unit
GBRF	GB Railfreight
GNR	Great Northern Railway
GWR	Great Western Railway
HST	High Speed Train (InterCity 125)
LMS	London, Midland & Scottish Railway
LNER	London & North Eastern Railway
LNWR	London & North Western Railway
LYR	Lancashire & Yorkshire Railway
MLS	Manchester Locomotive Society
MR	Midland Railway
NER	North Eastern Railway
PTE	Passenger Transport Executive
RCTS	Railway Correspondence & Travel Society
ROF	Royal Ordnance Factory
SLS	Stephenson Locomotive Society
TOC	Train Operating Company
TUCC	Transport Users Consultative Committee
UDC	Urban District Council
WYMCC	West Yorkshire Metropolitan County Council

ACKNOWLEDGEMENTS AND AUTHOR'S NOTE

Many of the images used in this book are drawn from the work of photographers whose work is now in the care of the Online Transport Archive. Other images have come from the Transport Treasury collection and I'm very grateful for the assistance of Andrew Royle in tracking these down and in forwarding on the scans. My very good friend Gavin Morrison has also been a great help in providing images from his extensive collection. The maps that were prepared for the book were produced by another very good friend, Paul Smith. My sincere thanks to all.

This book is not designed to be a detailed history of the railways of the area; there is little purpose in trying to repeat the comprehensive accounts of the growth and evolution of the system that have previously published. These works are listed in the bibliography. It is rather an exploration, largely through photographs with detailed captions, of the railway scene in the Bradford and Leeds area – effectively the region encompassed by Halifax, Brighouse, Normanton, Wetherby, Ilkley and Skipton – and how it has changed over the past 70 years.

My father was born in Halifax and grew up in a house literally across the road from West Vale station. He was three when passenger services on the branch ceased but he remembered the freight traffic heading to and from Stainland; indeed, his earliest transport photographs – sadly not of a quality for reproduction – were of freight trains at West Vale in the late 1930s. I spent my childhood in Bradford and have vivid memories of the old Exchange station and how bleak it was in its later years. Standing on the platforms ends watching the trains ascending and descending the gradient from Mill Lane Junction was always an impressive sight as even the most powerful locomotives could struggle to depart from the station at times.

Peter Waller,
Shrewsbury,
October 2023

INTRODUCTION

As one of the cradles of the Industrial Revolution, the area of Bradford and Leeds was amongst those that entered the railway age relatively early. At the western extremity of the great Yorkshire coalfield, where the coal was closest to the surface and thus most easily mined (the number of unknown bell pits found in the late 1980s when the Low Moor site was being prepared for the development of the West Yorkshire Transport Museum is testament to the early – and unrecorded – mining

A map showing the area covered in the book.

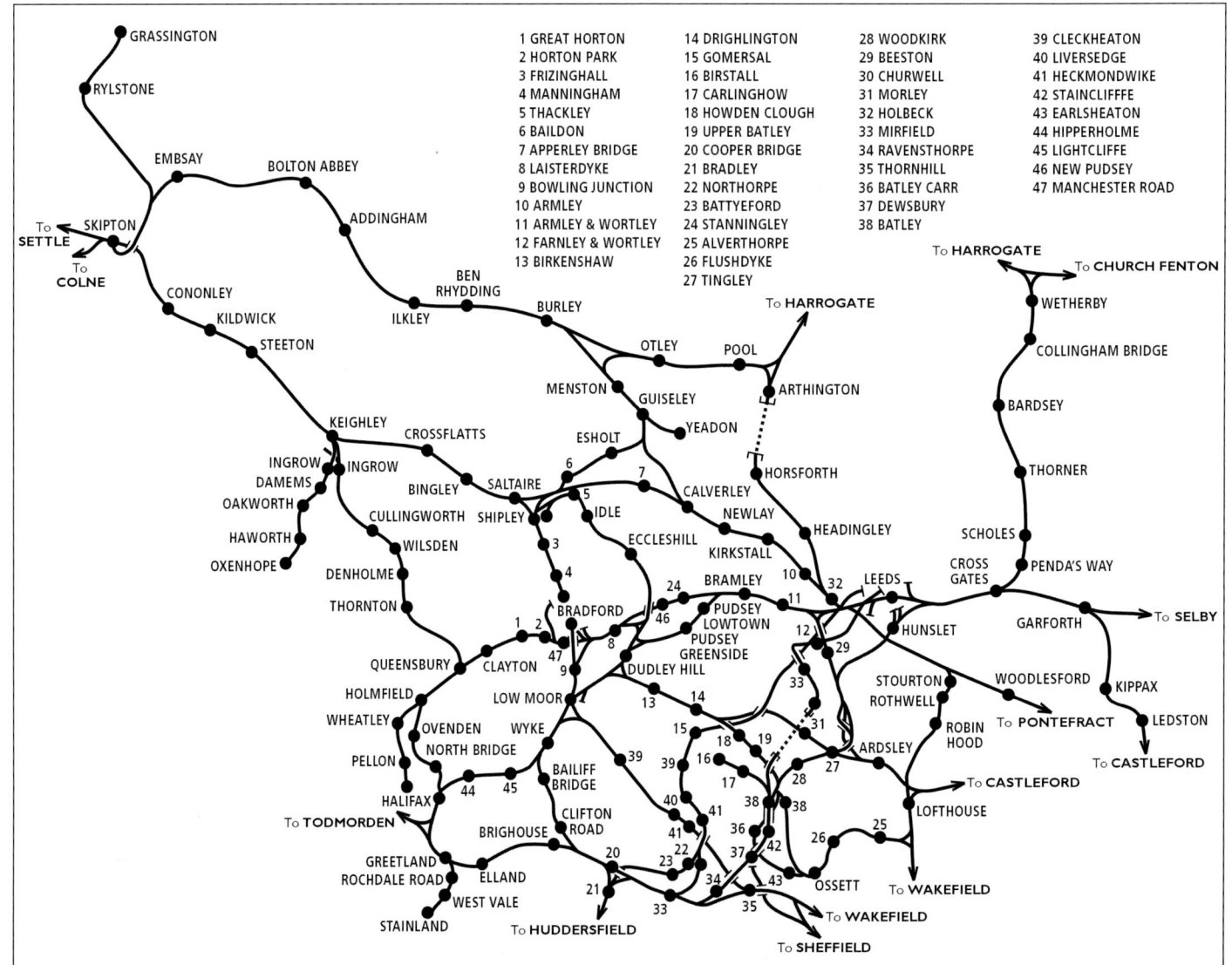

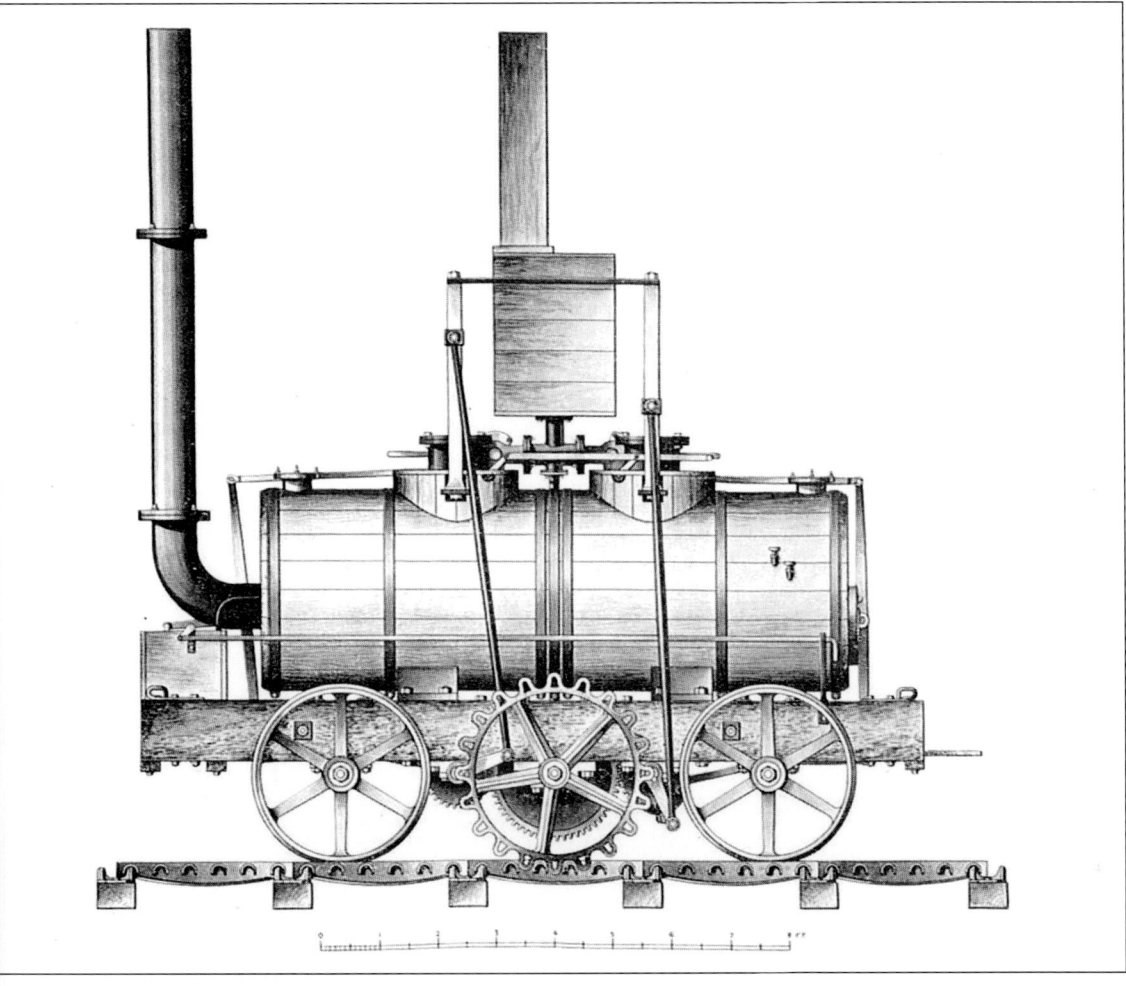

The first line to operate a steam locomotive commercially was the Middleton Railway in Leeds; John Blenkinsop, the colliery manager, patented toothed track in 1811 and engaged Matthew Murray to design and build a steam locomotive – *Salamanca* seen here in an engraving produced in 1829 – that incorporated the high pressure steam system patented by Richard Trevithick, for which Murray paid a royalty, alongside the double cylinder that Murray invented. The combination of toothed track and steam locomotive was a considerable success, with the latter capable of hauling nearly 20 times its own weight. Three similar locomotives were constructed with one being sent to the north-east where it influenced the design of George Stephenson's much less successful *Blücher*. The last of the Middleton locomotives operated in the mid-1830s. *The Mechanic's Magazine*

industry of the area), a significant number of collieries were developed from the early eighteenth century.

It was to serve the collieries south-west of Leeds that one of the most significant events of the early railway age took place. Although there had been a number of early wagonways serving mines and other industrial sites, none of these had been built with any form of statutory protection. Charles Brandling, the owner of the land on which the Middleton Colliery was developed, obtained on 9 June 1758 the first Act of Parliament to sanction the construction of a railway line. The preamble to the document read, 'An Act for Establishing Agreements made between Charles Brandling, Esquire, and other Persons, Proprietors of Lands, for laying down a Waggon-Way, in order for the better supplying the Town and Neighbourhood of Leeds, in the County of York, with Coals.' One of the local newspapers reported, on 26 September 1758:

> On Wednesday last the first Waggon Load of Coals was brought from the Pits of Charles Brandling, Esq, down to the new Road to his Staith near the Bridge in this Town, agreeable to the Act of Parliament passed last Sessions. … On this Occasion the Bells were set a ringing, the Cannons of our FORT fired, and a general Joy appear'd in every Face.

During the later eighteenth century the network served by the Middleton Railway expanded as more collieries were linked to it, with the result that by the end of the first decade of the nineteenth century, the system extended over more than four miles.

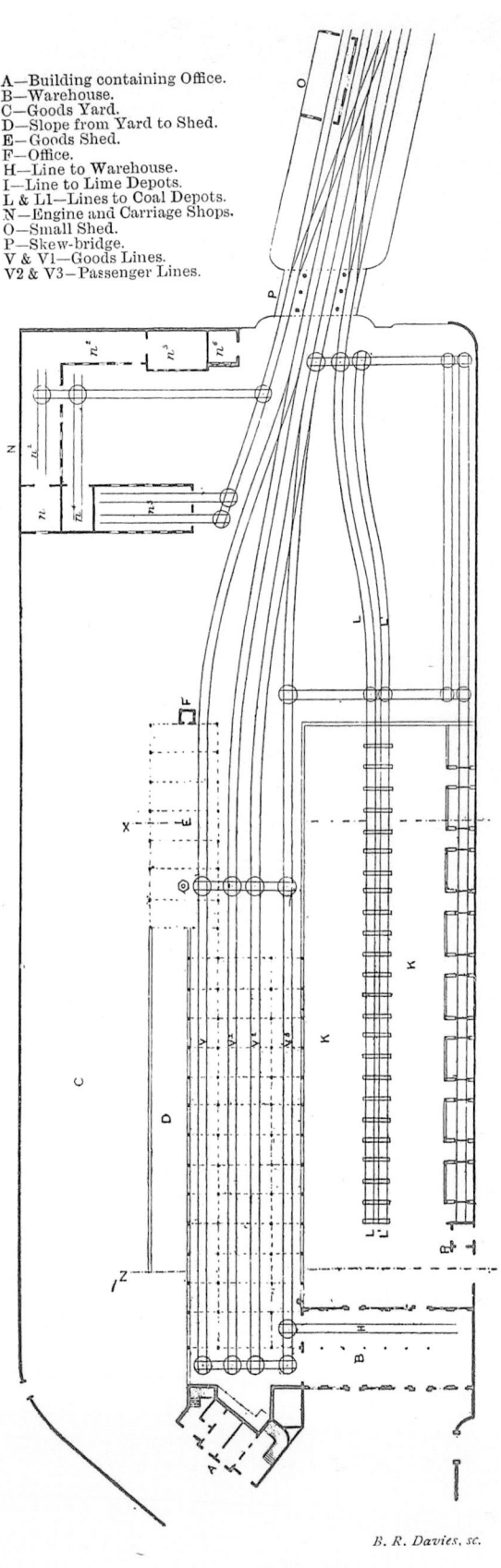

A—Building containing Office.
B—Warehouse.
C—Goods Yard.
D—Slope from Yard to Shed.
E—Goods Shed.
F—Office.
H—Line to Warehouse.
I—Line to Lime Depots.
L & L1—Lines to Coal Depots.
N—Engine and Carriage Shops.
O—Small Shed.
P—Skew-bridge.
V & V1—Goods Lines.
V2 & V3—Passenger Lines.

B. R. Davies, sc.

LEEDS STATION PLAN.

The track plan of the original Leeds & Selby terminus at Marsh Lane depicted in about 1842. The station originally opened on 22 September 1834 and the facilities when recorded here – viewed with west at the bottom and east to the top – accommodated both passenger and freight traffic. The site of the original station was redeveloped into a six-storey grain warehouse – destroyed by fire in the 1970s – that was designed by Thomas Prosser. The passenger station was closed on 1 April 1869 when a new station – on the line opened that day to serve Leeds New station – was opened. The new Marsh Lane station was closed on 15 September 1958.

The Middleton Railway was also the location of the first commercial application of steam traction. In 1811, conscious that any engine light enough to operate over the cast-iron rails would lack the adhesion required to haul a heavy train over steep gradients, John Blenkinsop, the mine's manager, patented a form of toothed rail. He approached the Newcastle-born engineer Mathew Murray, who had moved to Leeds at the age of 24 in 1789 and who – after working with the flax manufacturer John Marshall for a number of years – established with David Wood (later joined by James Fenton and William Lister) a factory at Holbeck to produce machinery, to design and build a steam locomotive. The result was *Salamanca*, the world's first twin-cylinder steam locomotive. The success of the new locomotive resulted in a further three being constructed, two of which operated on the Middleton Railway; the last of the trio survived in service until 1835.

By the start of the 1820s, the first proposals for major railways were being developed and, in 1824, the Leeds & Hull Railroad Co was established in Leeds. There were a number of factors in the development of many of these early railways; manufacturing industry was growing exponentially, and, with that, there was a growing need for the movement of raw materials and finished goods. The canal network was still growing but lacked capacity and competition (given the appalling state of the contemporary roads), with the result that manufacturers felt they needed an alternative means of shipping to and from the country's major ports. For Leeds, the obvious target was the port of Hull; however, the stock market crash of 1825 delayed the project whilst, in the meantime, the development of Goole as a port, following the opening of the eastern extension of the Aire & Calder Navigation (the Knottingley & Goole Canal) in 1826, offered an alternative destination.

The Leeds & Selby Railway was formed on 20 March 1829, with the new line surveyed by James Walker the same year. Construction of the line was authorised in an Act of 29 May 1830, which was passed despite opposition from the Aire & Calder Navigation. Work commenced on the line's construction on 1 October 1830; the work included the 700-yard-long tunnel

The exterior of Leeds Wellington station as recorded in an article published in *The Railway Magazine* in 1903. The station had originally opened on 1 July 1846 – largely replacing the Midland Railway's original terminus at Hunslet Lane which had opened originally on 1 July 1840 and was finally closed on 1 March 1851 – as a temporary station; the permanent station at the site opened on 1 October 1850. The building on the extreme left of the view is the original Queens Hotel; this had been built for the MR and opened in the 1860s. The old hotel was demolished in 1935 and replaced by the current hotel. *Author's Collection*

under Richmond Hill, which obviated the need for the planned inclined planes that had been a feature of George Stephenson's proposals for the earlier Leeds & Hull line. On 22 September 1834, following the completion of one line, the inaugural train departed from Marsh Lane station in Leeds; hauled by *Nelson*, one of four 2-2-0s supplied to the new railway by Edward Bury & Co for its opening; this train struggled to climb the gradient at Richmond Hill Tunnel. However, the journey to Selby was eventually completed – in about two-and-a-half hours – before returning to Leeds; the trip back was considerably less time-consuming, taking only sixty-five minutes. Initially, the line carried only passenger traffic; it was only after 15 December 1834, with the completion of the second running line, that freight traffic was also transported.

The completion of the Leeds & Selby marked the start of the creation of the network that came to serve the area. It was followed by the opening of the York & North Midland Railway; this was opened from York to Milford on 29 May 1839; opened at the same time was a curve from the north that gave trains from York access towards Selby. The York & North Midland was extended south to Burton Salmon on 11 May 1840. A second curve from Milford to join the Leeds & Selby was opened on 9 November 1840; this provided trains from the south with direct access to Selby. The York & North Midland was one of the lines controlled by the 'Railway King', George Hudson. Another Hudson controlled line, the North Midland, was opened from Masborough (Rotherham) through to Leeds via Normanton on 1 July 1840. The final section of the York & North Midland line – from Burton Salmon to connect with the North Midland Railway north of Normanton – opened on 30 June 1840.

The next of these pioneering railways to reach the area was the Manchester & Leeds Railway; although the first proposals for a trans-Pennine line from Manchester to Leeds dated to the mid-1820s, it was not until 4 July 1836 that the Act authorising construction of the route received the Royal Assent. Work commenced on 18 August 1837 on the line's construction. The section from Manchester to Littleborough opened on 3 July 1839; the chief booking clerk for the railway at Manchester was Thomas Edmondson, who invented the machine that bore his name for the printing of railway tickets. The section from Normanton, where a connection was made with the North Midland Railway in order to gain access to the latter's Leeds station on Hunslet Lane, opened on 5 November 1840. It was not until 31 December 1840, with the opening of the line through the 2,285-yard-long Summit Tunnel, that through services between Leeds and Manchester commenced.

The year in which these pioneering lines were completed marked the start of a decade in which the development of the railway network was dramatic. Places like Bradford and Huddersfield were to receive their first railways and, despite the setback caused by the loss of confidence caused by the collapse after the 'Railway Mania' later in the decade, by 1850, places such as Dewsbury, Harrogate, Skipton and Wetherby had all been added to the railway map. Over the next five decades the network was to expand further, with myriad – often competing – lines being constructed both to take railways into areas not already well served – such as the area bounded by Bradford, Halifax and Keighley – and increase capacity – as with the case of the LNWR's 'Leeds New' line – where it was impractical to improve the existing route. Even as late as the period immediately prior to the First World War there were proposals – only partially completed – by the MR to construct a further main line through the Spen Valley in order to reduce the distance on its main line from London to Carlisle (it was in competition with the shorter East Coast

The Railway Magazine of 1903 also had this view of the concourse area at Leeds Wellington; this was destined to be removed and replaced by the Grade II listed concourse designed by William Henry Hamlyn, the LMS's chief architect, and completed in 1938. *Author's Collection*

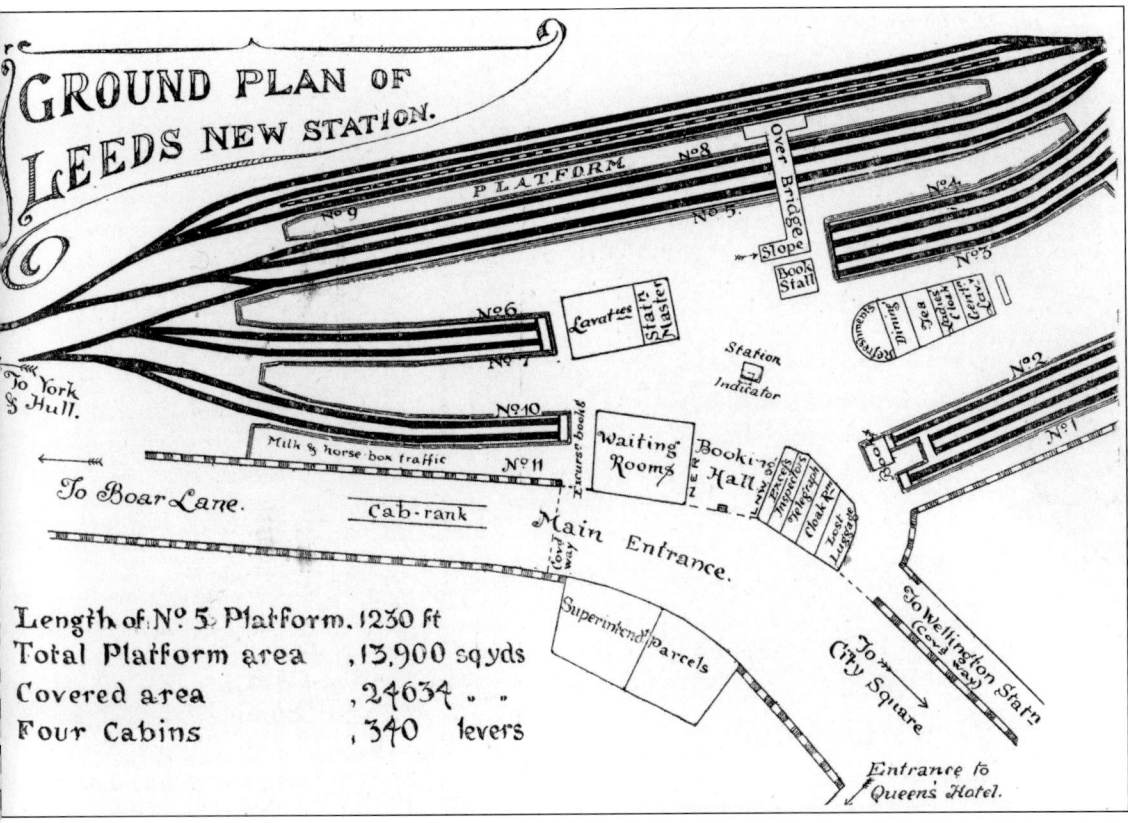

The track plan of Leeds New station as recorded in an article published in *The Railway Magazine* in 1904. The station, which was to provide accommodation for services operated by the NER and LNWR (and resulted in services being transferred from Leeds Central), opened on 1 April 1869. *Author's Collection*

and West Coast main lines for the lucrative Anglo-Scottish traffic). In the event, the war and the post-war economic decline put paid to these final grandiose plans; had they happened, however, Bradford would have been placed on a main line and the debate about its lack of decent rail connections – which continues at the time of writing – would have been a thing of the past.

By the end of the nineteenth century there were five dominant railway companies in the Bradford and Leeds area – the GNR, LNWR, LYR, MR and NER – along with a number of joint lines. The chronology of the individual railway companies and of the routes that they constructed is detailed in the appendix. By 1900, however, the first threat to the hegemony of the railway locally was already starting to appeal. As the area's towns and cities grew, so the need for improved local public transport developed. In both Leeds and Bradford, tramway systems – originally horse and later steam – developed but it was only after the introduction of electric traction that the tramway networks expanded and started to provide real competition for the often ill-located railway stations. Whilst, for example, the GNR could provide a service to stations such as Thackley, Idle and Eccleshill on its circuitous line from Laisterdyke to Shipley, the frequency of service and the fares charged gave Bradford City Tramways a competitive edge, whilst Queensbury station was more than a mile from the centre of the community and some 400ft below it – the trams, from both Bradford and Halifax (there was no through service due to a six-inch difference in gauge), terminated in the centre. The first railway passenger service to succumb largely as a result of tramway competition was the short-lived branch from Stourton to Rothwell that lasted less than a year in 1904. Tramway competition was one factor in the closure of Manchester Road station in Bradford in 1916.

By this date, however, Britain was at war; legislation passed in the First World War permitted the railways to close stations and lines. A number of passenger

Halifax station recorded from the south as illustrated in an article in *The Railway Magazine* in 1905; the station here originally opened with the line through towards Bradford on 7 August 1850. Originally provided with temporary wooden buildings, permanent buildings to the design of Thomas Butterworth, in a neo-classical style, were opened on 23 June 1855. The station underwent significant rebuilding between 1884 and 1886 in order to accommodate the services operated into the town by the GNR on the route via Queensbury. The station as rebuilt had separate accommodation for the two railway companies: that for the LYR on the east and that for the GNR on the west. *Author's Collection*

services – including the Halifax High Level line to St Paul's, the LNWR branch from Batley to Birstall and the MR cut-off route from Royston Junction to Thornhill Junction – and stations – such as Bailiff Bridge, Bowling Junction and a number on the 'Leeds New' line – closed as a result; most were never reopened after the war. Few lines were to close completely during this era, however, although one route that did was the two-mile GNR branch from Dudley Hill to Low Moor; this lost its passenger services on 31 August 1914 and was closed completely on 1 October 1917. The GNR goods yard at Low Moor continued to receive traffic until May 1933, with traffic being routed via the LYR line.

The return to peace did not bring a return to the glory days of Edwardian prosperity; the railways faced a new challenge with the rise of unfettered motor transport competition – it was not until 1930 that bus operations were regulated and the availability of ex-military road vehicles saw the growth of new bus companies; these were a threat not only to the railways but also to the established tramways. The immediate post-war years represented an era of industrial stagnation with many of the traditional industries struggling to survive; there was also the Grouping of the railway companies in 1923. The LYR and LNWR had already merged the previous year but now the enlarged LNWR merged with the MR and the GNR with the NER to form the LMS and LNER respectively; joint lines – such as the Otley & Ilkley – retained this status. As the economic conditions deteriorated further and as competition continued, so the process of gradual contraction continued. The ex-LYR branch from Greetland to Stainland lost its passenger services on 23 September 1929 and the ex-GNR line from Laisterdyke to Shipley followed on 2 February 1931. The lines remained open for freight traffic, however.

The local railway network passed virtually intact to BR on 1 January 1948; the railways of Bradford and Leeds fell into three of the new BR regions. The ex-LMS lines initially passed to the London Midland Region; the ex-NER lines of the LNER to the North Eastern Region and the ex-GNR lines to the Eastern. Over the next decade there were some revisions to the regional boundaries, but the most significant change came in 1967 with

The arrival of the tramcar represented a serious challenge to the established railways; often operating in direct competition, the tramcars offered a more frequent and convenient service often by a less circuitous route. Halifax Corporation operated an extensive network of 3ft 6in electric trams between 1898 and 1939 with a system that extended – at its peak – over more than 39 route miles. A number of these routes – to, for example, Hebden Bridge, Queensbury and Stainland – operated parallel to the railways. Pictured on Stainland Road, West Vale, with the LYR viaduct that carried the Stainland branch across the valley in the background, is Halifax tram No 52 about to head inbound to Halifax town centre. The tramway extension through to Stainland opened on 14 May 1921; the ex-LYR Stainland branch lost its passenger services less than a decade later. *Geoff D. Smith Collection/Online Transport Archive*

- Bradford to Huddersfield via Mirfield (local services) – this service was withdrawn on 14 June 1965.
- Bradford to Huddersfield via Halifax (local services) – this service was withdrawn on 14 June 1965; however, a Bradford to Huddersfield service was restored on 28 May 2000 following the reopening of Brighouse station.
- Leeds City/Bradford Forster Square to Skipton via Ilkley – the service was retained as far as Ilkley but closed beyond there to Skipton on 22 March 1965. Although the route from Arthington through Otley was not specifically identified separately, the appendix in the report included the stations of Arthington, Pool-in-Wharfedale and Otley as potential casualties and the service was also withdrawn on 22 March 1965 although the last passenger service over the route – a special from Preston to York – travelled via Skipton, Ilkley and Otley.
- Leeds City/Bradford Forster Square to Skipton via Keighley (local services) – service not withdrawn.
- Leeds City to Bradford Forster Square via Shipley (local services) – this service was withdrawn on 22 March 1965.
- Leeds City to Sheffield Midland via Cudworth (local services) – this service was withdrawn on 1 January 1968; main-line non-stop services were restored via the route on 6 May 1974 but transferred away from the section south of Goose Hill Junction on 31 August 1985.
- Leeds City to Micklefield via Cross Gates (local services) – service not withdrawn.
- Leeds City to Harrogate via Wetherby – service withdrawn 6 January 1964.
- Wetherby to Church Fenton – service withdrawn 6 January 1964.

The normal procedure for any proposed closure was that a closure notice was erected at the affected stations inviting objections. In the case of the lines from Bradford and Leeds to Skipton via Otley and Ilkley and the surviving local stations for the local services between Leeds and Bradford Forster Square, the closure notices were put up in December 1963 inviting objections. The next stage was the hearing of the TUCC, the body that met to determine whether what hardships might arise as a result of the line's closure and to advise the Minister of Transport on this. In the case of the lines through Ilkley, the TUCC met in Leeds in May 1964. There had been particularly strenuous opposition from commuters on the stations served by the trains from Ilkley to Leeds and Bradford; these had formed the Ilkley Railway Supporters Association to campaign for the line's retention. The attitude of the local authorities was less consistent; the town council in Otley did not oppose closure, arguing that the local buses were adequate, whilst councillors in Ilkley were more favourable to the line's retention. In the event, the decision later that year was to confirm the closure of the sections between Ilkley and Skipton and between Guiseley and Arthington but that a decision on the remaining sections would be deferred.

This process was replicated on all the lines proposed for closure across the country; some were reprieved entirely, some survived partially and other were closed. During the course of the 1960s, the passenger network in the Bradford and Leeds area shrank considerably. The same was equally true of the once extensive freight network. Historically, the railway had had a position as 'common carrier'; this meant that the industry had to carry any freight offered to them at a nationally agreed charge. The result

Bradford Forster Square station at its apogee during the early years of the twentieth century. In 1890, work on rebuilding Forster Square station to the designs of the MR's architect Charles Trubshaw was completed. The work included both a passenger and goods station as well as the new Midland Hotel. It was in the lobby of the Midland Hotel that the noted Victorian actor Sir Henry Irving died after suffering a stroke on 13 October 1905 when on the stage at the city's Theatre Royal on Manningham Lane. This undated photograph must be roughly contemporaneous with the death of Irving as tram No 228 was new in 1903 and was withdrawn in 1916. Forster Square itself was laid out in the late nineteenth century and named in honour of William Edward Forster, who had been MP for Bradford and had served in Gladstone's governments, following his death in 1886. The hotel – which is now Grade II listed – was restored in 1993 following privatisation. *Geoff D. Smith Collection/Online Transport Archive*

Viewed looking towards the east, this busy scene of City Square in Leeds sees trams heading to and from Boar Lane with, on the right, the imposing bulk of the then new Queens Hotel. Following the demolition of the original hotel in 1935, a new hotel – designed alongside the new concourse at Wellington station by William Henry Hamlyn but incorporating interior designs and art deco features produced by William Curtis Green – was opened by Lord Harewood and his wife, the then Princess Royal, on 12 November 1937. Taken over by British Transport Hotels in 1948 and privatised during the 1980s, the building is still a hotel and is now Grade II listed. *Geoff D. Smith Collection/Online Transport Archive*

of this was that a network of goods yards, capable of handling a variety of traffic, survived even if the adjacent passenger station was closed. There had been some attempt at rationalisation and concentration but the traditional wagon-load traffic with pick-up freights and much shunting at yards was still widespread in the industry. The Transport Acts of 1953 and 1962 effectively removed this responsibility from the railway industry – reflecting the post-war drift towards increasing use of lorries and commercial vehicles – and Dr Beeching's analysis demonstrated that the only profitable future for the railway in freight traffic was in block trains and containerisation. The result of this was that freight facilities were increasingly withdrawn from stations – a process accelerated by the decline in many of the traditional industries – and on a monthly basis the pages of *The Railway Magazine* recorded these losses. By the late 1970s, the once extensive network of lines retained for freight traffic only had diminished significantly and by the end of that decade there remained a small number of stations still handling freight and often this was little more than domestic coal. Thus, for example, freight facilities were withdrawn from Dudley Hill in August 1979 and Shipley in September 1980.

By the 1980s the passenger railway network was largely stabilised; there were still odd closures – most notably in the West Yorkshire area of the Clayton West branch and the ex-MR main line south from Normanton (with traffic diverted to alternative routes) – but the massive reduction wrought during the 1960s was a thing of the past. The first indication that there might be a revival came with the reopening of Baildon station on 5 January 1973. West Yorkshire PTE was one of the more pro-rail of the Metropolitan County Councils and prior to its abolition had undertaken major strategic review of the railway services within the county; this came up with a range of options, many of which included the provision of reopened stations and lines. The result has been over the past 30 years that there has been a slight increase in the passenger network – most notably with services restored to the ex-Manchester & Leeds railway line from Halifax to Huddersfield via Brighouse – and with a number of stations either being reopened or new stations – as at Crossflatts – provided.

For the past three decades, following the passing of the 1993 Railways Act, passenger services have been franchised; a number of franchisees have provided services in succession to InterCity and Regional Railways locally with varying degrees of success. Until the recent Covid-19 pandemic, railway usage both locally and nationally was increasing; on services in Bradford and Leeds this had resulted in increasing congestion and significant investment being made – for example in the further rebuilding of Leeds station – to increase capacity. Whether the short-term loss of traffic that has resulted

Although the BR period – and in particular the period after the Beeching Report of March 1963 – is generally regarded as the period when the railway network contracted; after the Grouping in 1923, the opportunity was taken to eliminate duplication. One of the casualties of this period was the passenger service to the former LYR terminus Dewsbury Market Place, which was closed on 1 December 1930 but retained for goods traffic – as seen here – until final closure more than 30 years later. The station had opened on 1 April 1867 and gained the 'Market Place' suffix on 2 June 1924. *Transport Treasury*

20 • THE RAILWAYS OF BRADFORD AND LEEDS – THEIR HISTORY AND DEVELOPMENT

Right: **With its** huge natural coal reserves, Britain was slow to explore alternative means of traction. Limited use of diesel operation began before the Second World War, but it was only in the 1950s that dieselisation moved rapidly forwards. The introduction of DMUs provided two benefits: they were more economical to operate and were perceived as cleaner than their steam alternatives. However, in the harsh economic reality of the post-war years, DMUs were often introduced too late and services that they might have helped to save were still lost. The first major investment into DMUs went into the Derby Lightweight units – one of which (with Driving Motor Composite No E79511) is pictured here at Bradford Exchange station. Bradford Hammerton Street depot was one of the first to be modified to cater for the mass introduction of this type of train. *Transport Treasury*

Below: **From the** 1960s onwards, the railways witnessed a considerable amount of destruction – an inevitable consequence of line and station closures. Not all demolition was, however, for negative reasons as this view of Leeds City taken in the early 1960s evinces. Here, it was a case of removing the old Leeds New station in order to facilitate the construction of the new Leeds City station. More than three decades after that work was completed, a new project saw the station further rebuilt to cater for the growth of passenger traffic. *Transport Treasury*

INTRODUCTION • 21

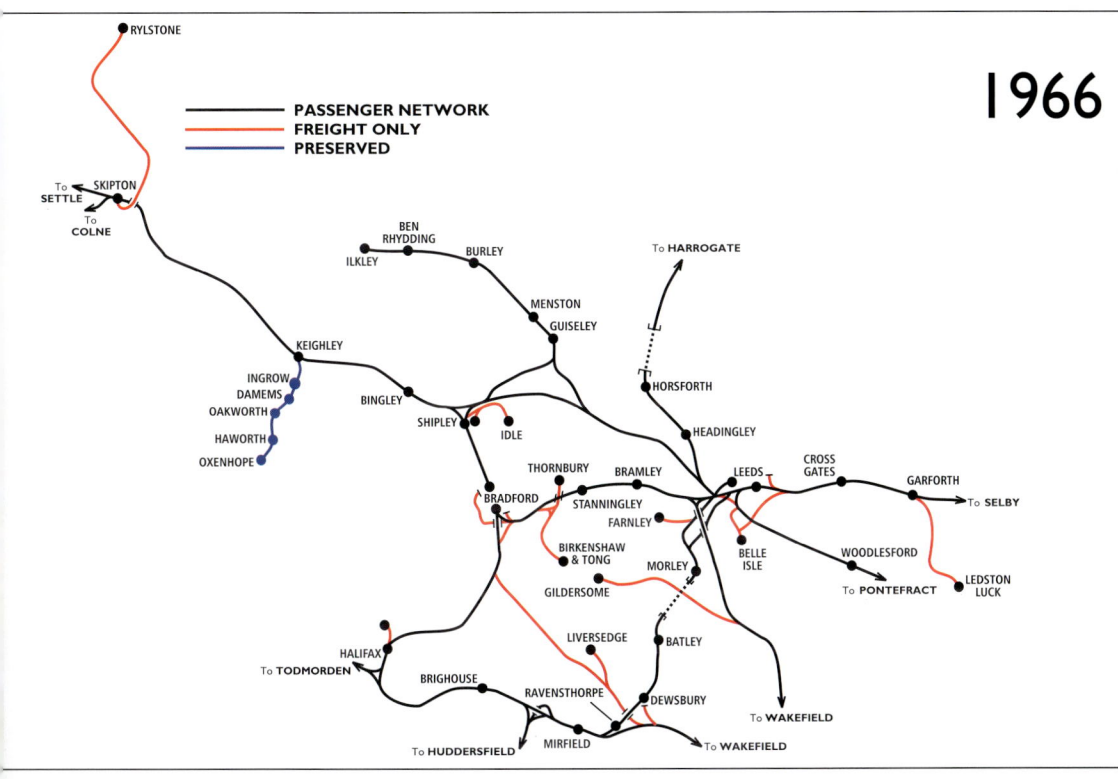

The evolution of the railway network over the past 60 years; this map portrays the railways of Bradford and Leeds at the start of 1966 by which date the majority of the closures envisaged by the Beeching report of March 1963 had been undertaken.

from the pandemic turns into a longer term drift away from the railways is uncertain but undoubtedly future plans will come under closer scrutiny. For freight – again privatised in the 1990s – the potential for traffic growth seems greater as concerns over the environment and congestion grows.

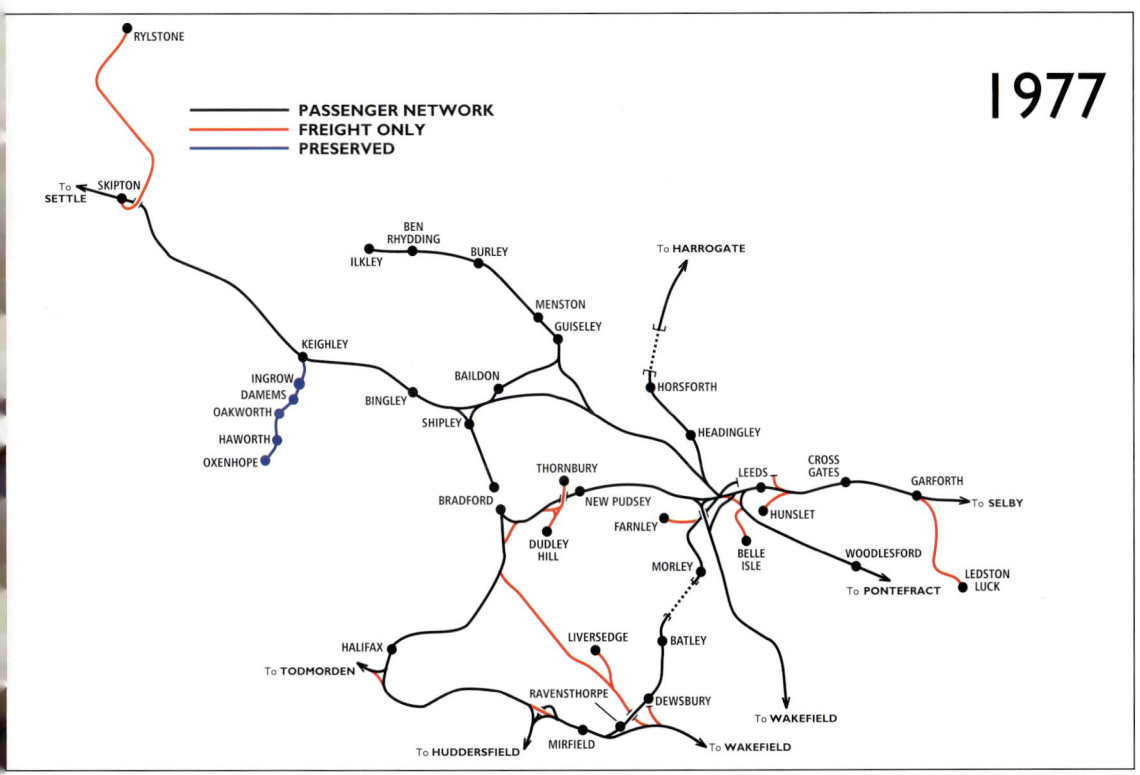

A decade later in 1977 although the passenger network has not shrunk significantly – other than the loss of the services over the ex-Manchester & Leeds main line between Sowerby Bridge and Mirfield, the reduction in the freight-only network is notable as a number of the major freight facilities – such as Adolphus Street and City Road in Bradford – are closed over the previous decade.

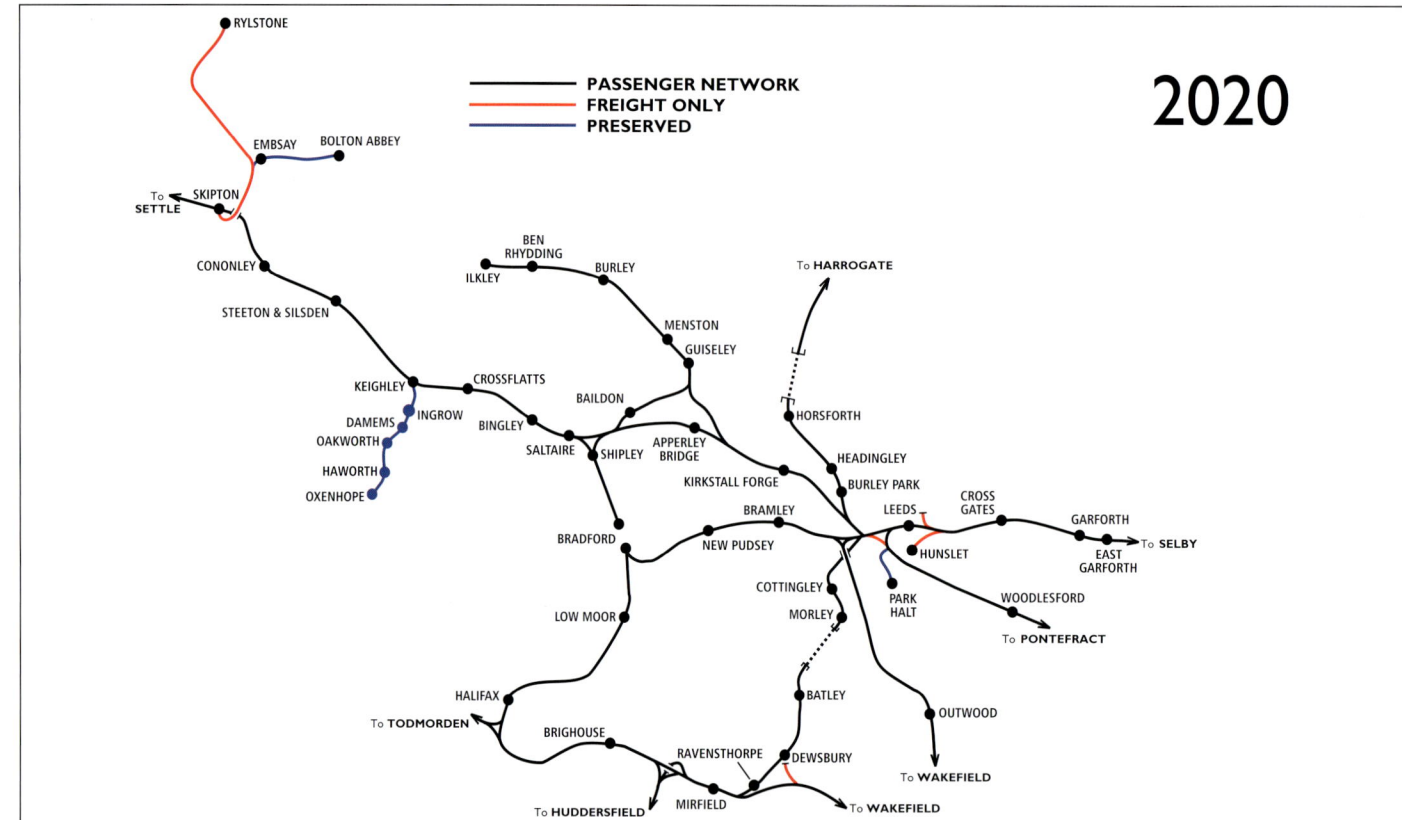

Above: **The freight-only** lines have virtually disappeared by 2020 but the expansion of the passenger network and in particular the opening or reopening of stations is notable. The previous thirty years had witnessed a virtually relentless increase in the level of passenger traffic – with often overcrowding as a result – but this came to a halt at the start of 2020 as the country faced the arrival of the Coronavirus pandemic. At the time of writing, the economy is gradually being reopened and most of the restrictions imposed as a result of the virus have been lifted; what the long-term effects of the virus on the railway industry both locally and nationally are it is too early to say but it seems likely that the forecasts of traffic growth may well have to be revised downwards and, with that, the viability of a number of proposed schemes may well be called into question. Conversely, the challenges set by climate change and the need to become a carbon neutral economy by the middle of the century may well result in greater investment in certain types of public transport. In early 2021, the West Yorkshire Combined Authority launched proposals for a mass transit scheme that would include light-rail and tram-trains. There is still no certainty that these proposals will get beyond the drawing board – the history of public transport provision in the area is littered with plans that came to nothing (including the MR's scheme for the Bradford cut-off line and, most recently, the Leeds section of the High Speed network) – but, if these do develop (as has been the case in Manchester), then there is every likelihood that a new edition of this book in a decade's time might well show a considerably enlarged local rail network.

Opposite above: The Rail Blue livery that dominated the railway landscape for some two decades was launched in 1964; typical of the scene in the 1970s is this view at Leeds taken on 5 July 1977 that sees Class 47/4 No 47461 arriving from the west as Class 45 No 45003 awaits the road to depart light engine towards Holbeck. In the background, the DMU is in the reversed livery that was carried by refurbished units; it also carries the stylised White Rose symbol, created from a circular pattern of interlinked 'WY's, that was a feature of locally based DMU stock for a period. *Author*

Opposite below: For the past three decades, since the privatisation of the railway industry following on from the Railway Act of 1993, operation of passenger services has been in the hands of various franchisees; some have had some success, others have been a litany of failure and this has been reflected in the services operated on the routes serving Bradford and Leeds. In this view – taken on 13 March 2020 just before the country entered the first national lockdown that resulted from the Covid-19 pandemic – Northern Class 153 No 153324 stands in Mirfield station with a service towards Huddersfield whilst a TransPennine express Class 185 passes through the station at speed heading for Manchester. At the time of writing, further change to the railway industry, partly the consequence of problems relating to the franchise process and partly a result of the changed environment post-Covid, means the creation of a new body – Great British Railway – to take over responsibility for the industry in 2023. *Gavin Morrison*

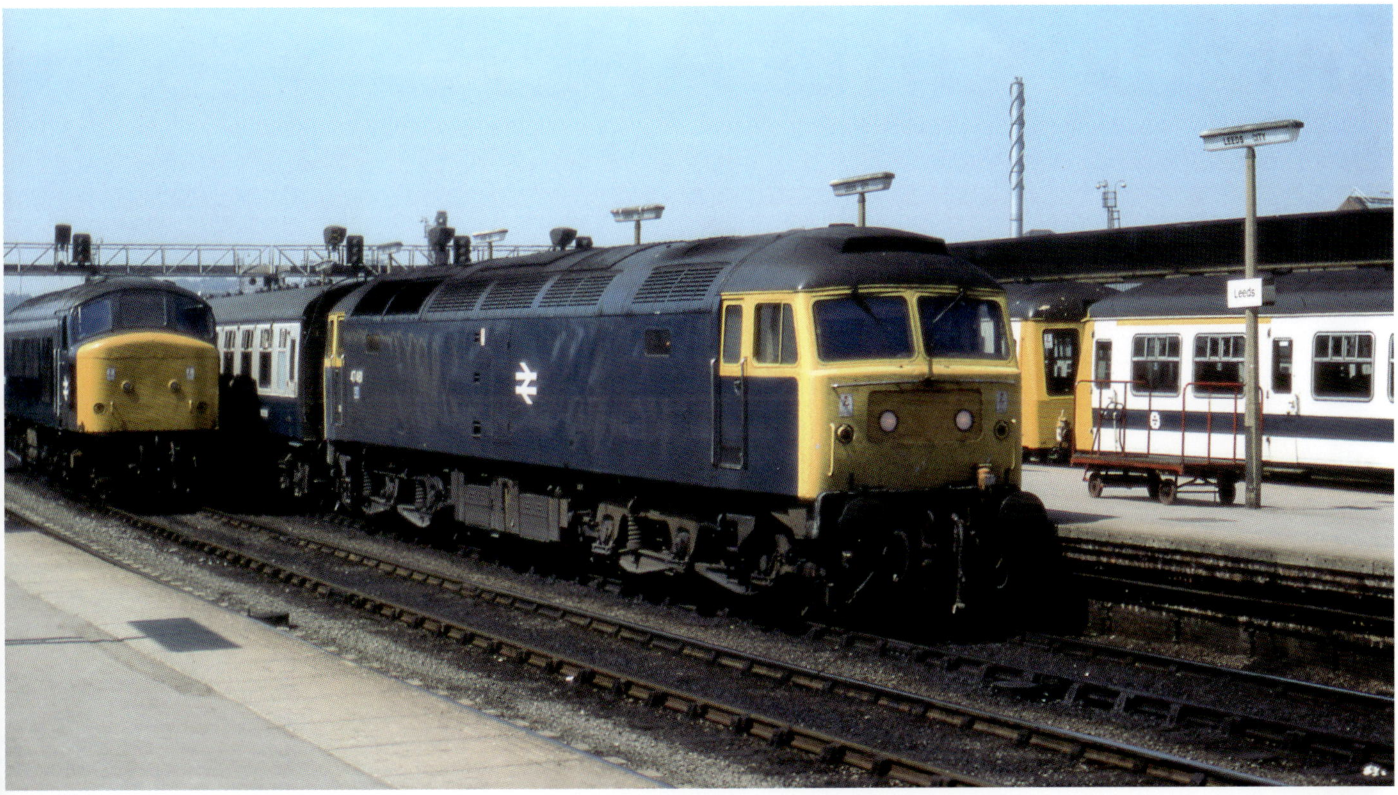
Leeds

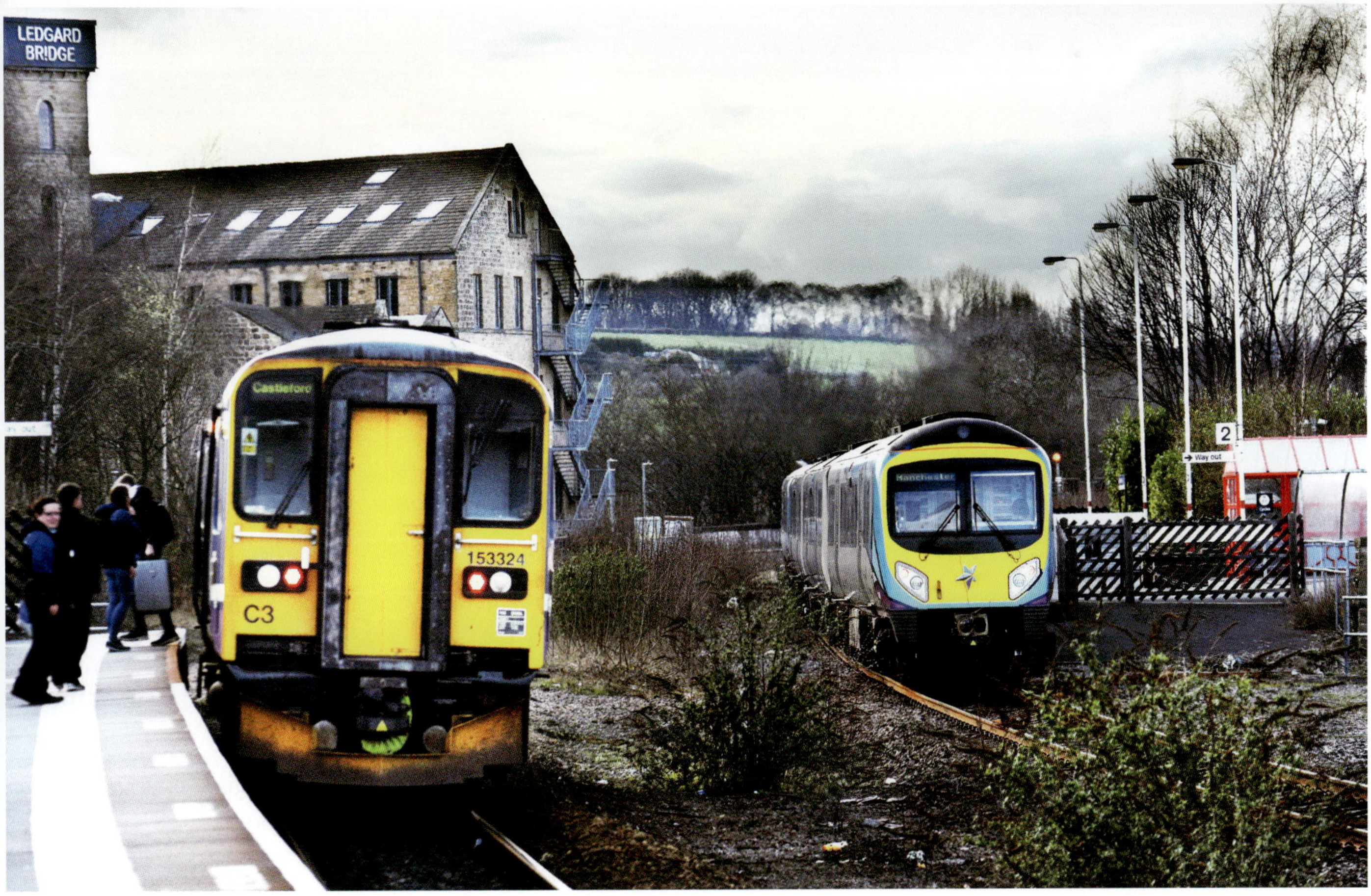
LEDGARD
BRIDGE
Castleford
153324
C3
Manchester
2
Way out

BRADFORD AREA

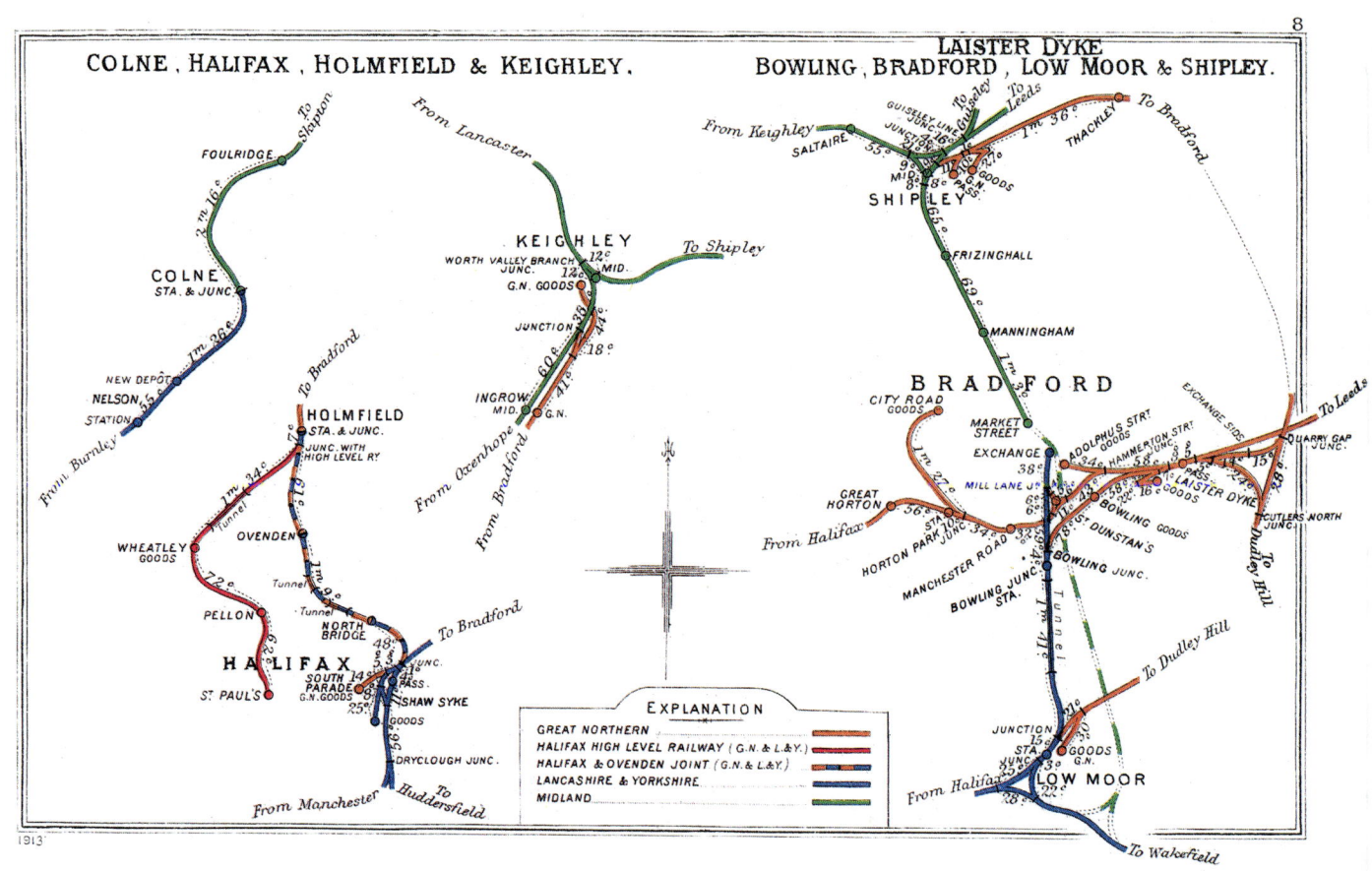

Above: **The Railway** Clearing House maps for Bradford, Halifax and Keighley for 1913. Of particular note are the route of the proposed MR extension from the Spen Valley via Low Moor to connect with the existing line and the fact that the MR's station in the city was still known as Market Street; it would not officially become Forster Square until after the First World War.

Opposite above: **St Dunstan's** station was provided with platforms on the lines towards Laisterdyke and towards Queensbury. Pictured entering the latter with a service from Bradford Exchange to Keighley on 30 July 1951 is Class N1 No 69474. The line from St Dunstan's to Thornton was authorised in 1871; it opened from St Dunstan's to Great Horton along with the freight-only branch to City Road to goods traffic in November 1876. Goods traffic was extended to Clayton on 9 July 1877 and to Thornton by April 1878; passenger traffic over the line to Thornton commenced on 14 October 1878. After the Second World War, traffic on the Queensbury triangle routes declined as private car ownership increased and, as a result, economies were sought. As a result, St Dunstan's station – along with that at Horton Park – was closed on 15 September 1952. St Dunstan's station was subsequently demolished, although trains travelling from Bradford Interchange to Leeds still pass through the site. Passenger traffic over the line to Thornton closed on 23 May 1955 but the line remained open for freight to City Road and to Great Horton until 28 July 1972 and 26 August 1972 respectively. *Tony Wickens/Online Transport Archive*

On 31 July 1951, Class N1 0-6-2T No 69572 pauses at Drighlington & Adwalton station with a service heading towards Ardsley. The line from Laisterdyke to Ardsley was opened to Gildersome to passenger traffic on 20 August 1856 and thence to Ardsley on 10 October 1857. The station – by now called simply Drighlington – lost its passenger services on 1 January 1962 although the route from Laisterdyke through to Ardsley remained open for through services until 4 July 1966. *Tony Wickens/Online Transport Archive*

Above: **Viewed** looking towards Bradford in 1953, a Class B1, No 61033 *Dibatag,* is seen approaching the station at Drighlington with a service heading towards Morley. The station opened on 20 August 1856 and was officially known as Drighlington & Adwalton until 12 June 1961 when it became simply Drighlington. Closed to passenger traffic on 1 January 1962, freight facilities were provided until 1 June 1964. The station was situated on the section of the Laisterdyke to Ardsley route that was closed completely on 31 October 1966. The trackbed and station site here have been swept away as a result of the construction of the Drighlington bypass. *Neville Stead Collection/Transport Treasury*

Opposite above: **The 'Queensbury Triangle'** timetable for the summer of 1947. *Author's Collection*

Opposite below: **Pictured with** a Bradford-bound service in Queensbury station is Class N1 No 69478. The station was remarkably inconvenient for the community it purported to serve, being some 400ft below the settlement and accessed only via a badly-illuminated footpath. Commercially this was to become a problem when both Bradford and Halifax commenced tramcar operation to Queensbury – the terminus there being the highest in Britain – and passengers no doubt increasingly preferred the tramcars and later the buses that the two corporations provided. Although the liner from St Dunstan's to Thornton opened to passenger traffic on 14 October 1878, it was not until 14 April 1879 that the station serving Queensbury was opened. Passenger services on the line to Holmfield and Halifax commenced on 1 December 1879. Freight traffic continued to be handled at the station after the closure to passenger traffic, with freight facilities not being withdrawn until 11 November 1963. Since the line's closure, the station and viaduct have been demolished and the site subjected to infill. The portals to both Clayton and Queensbury tunnels are still extant. Queensbury Tunnel is, at the time of writing, the subject of controversy, with its current owners – Highways England – seeking, on grounds of damage and lack of maintenance, to seal the tunnel off whilst local campaigners are keen to see the structure incorporated in a proposed cycleway and footpath that would make use of the ex-GNR trackbed. Sections of the trackbed have already been converted and there are plans to create a route that links Bradford with Keighley and Queensbury with Halifax. *Neville Stead/Transport Treasury*

Table 94 **BRADFORD, KEIGHLEY, and HALIFAX (via Queensbury)**

Week Days

Miles		p.m	a.m	a.m		a.m	a.m	a.m	a.m		p.m	p.m	a.m	
		A	B	A		A	A	B	A		B	A	B	
	1 London (King'sCross) dep	10J40	..	..	..	..	..	..	4 25	..	..	..	8R45	..
—	**Bradford** (Exchange)... dep	5a45	5 55	6 30	..	7 15	7 54	9 3	10 8	..	12 10	12 50	1p30	..
½	St. Dunstan's	5 49	5 58	6Y35	..	7 20	7 58	9 6	10 11	..	12 13	12 53	1Z36	..
1¼	Horton Park	5 53	6 3	6 39	..	7 24	8 2	9 10	10 15	..	12 17	12 57	1 40	..
2¼	Great Horton	5 56	6 7	6 42	..	7 27	8 5	9 13	10 18	..	12 20	1 0	1 43	..
3¼	Clayton	6 0	6 12	6 46	..	7 31	8 9	9 17	10 22	..	12 24	1 4	1 47	..
4¼	**Queensbury** arr	6 4	6 17	6 50	..	7 35	8 13	9 21	10 26	..	12 28	1 8	1 51	..
—	Queensbury dep	..	6 18	6N55	..	..	8N18	9 25	10N35	..	12 29	..	1 52	..
6¼	Thornton	..	6 22	6 59	..	..	8 22	9 29	10 39	..	12 33	..	1 56	..
7¼	Denholme	..	6 25	7 2	..	..	8 25	9 32	10 42	..	12 36	..	1 59	..
8¼	Wilsden	..	6 28	7 5	..	..	8 28	9 35	10 45	..	12 39	..	2 2	..
9¼	Cullingworth	..	6 32	7 9	..	..	8 31	9 38	10 48	..	12 42	..	2 5	..
12½	Ingrow	..	6 39	7 16	..	..	8 38	9 45	10 55	..	12 49	..	2 12	..
13½	Keighley arr	..	6 42	7 19	..	..	8 41	9 48	10 58	..	12 52	..	2 15	..
—	**Queensbury** dep	6 7	..	6 58	..	7 38	8 15	9N26	10 34	..	12N32	1 15	..	..
7	Holmfield	6 11	..	7 2	..	7 42	8 19	9 30	10 38	..	12 36	1 19	..	..
7¾	Ovenden	6 13	..	7 4	..	7 44	8 21	9 32	10 40	..	12 38	1 21	..	..
9	**Halifax** (North Bridge)	6 17	..	7 8	..	7 48	8 25	9 36	10 44	..	12 42	1 25	..	..
9¾	" (Old) arr	6 19	..	7 10	..	7 50	8 27	9 38	10 46	..	12 44	1 27	..	..

Miles		a.m	p.m	p.m.	p.m		p.m	p.m		p.m	p.m	p.m	p.m	Sundays		
		A	A	B	B		A	A			A	B	A			
	1 London (King'sCross) dep	10R10	..	12S35	..	..	1R10	1R25	..	..	4R 0	..	6RT5	..	..	..
—	**Bradford** (Exchange)... dep	3p30	4 38	5 15	5 45	..	6 0	6 46	..	..	9 15	10 27	1120	..	..	..
½	St. Dunstan's	3 34	4 41	5 18	5 48	..	..	6 49	..	..	9 19	..	..	..	..	..
1¼	Horton Park	3 38	4 45	5 22	5 52	..	6 6	6 53	..	..	..	..	..	..	..	..
2¼	Great Horton	3 41	4 48	5 25	5 55	..	6 9	6 56	..	..	9 25	10 34	..	..	..	..
3¼	Clayton	3 45	4 52	5 29	5 59	..	6 13	7 0	..	..	9 29	10 38	..	..	..	..
4¼	**Queensbury** arr	3 49	4 56	5 33	6 3	..	6 17	7 4	..	..	9 33	10 42	..	..	..	..
—	Queensbury dep	3N54	5N 0	5 37	6 7	..	..	..	..	..	..	10 49	..	..	..	..
6¼	Thornton	3 58	5 4	5 41	6 11	..	..	..	..	..	..	10 53	..	..	..	..
7¼	Denholme	4 1	5 7	5 44	6 14	..	..	..	..	..	..	10 56	..	..	..	..
8¼	Wilsden	4 4	5 10	5 47	6 17	..	..	..	..	..	..	10 59	..	..	..	..
9¼	Cullingworth	4 7	5 13	5 50	6 20	..	..	..	..	..	..	11 2	..	..	..	..
12½	Ingrow	4 14	5 20	5 57	6 27	..	..	..	..	..	..	11 9	..	..	..	..
13½	Keighley arr	4 17	5 23	6 0	6 30	..	..	..	..	..	..	11 12	..	..	..	..
—	**Queensbury** dep	3 52	5 4	5N37	..	..	6 18	7 8	..	8 8	9 37	10N51	..	..	..	..
7	Holmfield	3 56	5 10	5 41	..	..	6 22	7 12	..	8 12	9 41	10 55	..	..	..	..
7¾	Ovenden	3 58	5 12	5 43	..	..	6 24	7 14	..	8 14	9 43	10 57	..	..	..	..
9	**Halifax** (North Bridge)	4 2	5 16	5 47	..	..	6 28	7 18	..	8 18	9 47	11 1	..	..	..	..
9¾	" (Old) arr	4 4	5 18	5 49	..	..	6 30	7 20	..	8 20	9 49	11 3	1143	..	..	..

Week Days

Miles		a.m	a.m	a.m		a.m	a.m	a.m		p.m	p.m		p.m	
		B	B	B		A	A	B		A	B		A	
	Halifax (Old) dep	6N35	..	7N50	..	8 20	9 5	10N12	..	12 3	..	..	1 17	..
¾	" (North Bridge)	6 38	..	7 53	..	8 23	9 8	10 15	..	12 6	..	..	1 20	..
2	Ovenden	6 42	..	7 57	..	8 27	9 12	10 19	..	12 10	..	..	1 24	..
2¾	Holmfield	6 45	..	8 0	..	8 30	9 15	10 22	..	12 13	..	..	1 27	..
5	**Queensbury** arr	6 51	..	8 6	..	8 36	9 21	10 28	..	12 19	..	..	1 33	..
—	Mls **Keighley** dep	6 25	7 5	7 37	..	..	8N52	10 1	..	11N50	1242	..	..	..
—	1 Ingrow	6 31	7 11	7 43	..	..	8 58	10 7	..	11 56	1248	..	..	..
—	3¾ Cullingworth	6 39	7 19	7 51	..	..	9 6	10 15	..	12 4	1256	..	..	..
—	4¾ Wilsden	6 43	7 23	7 55	..	..	9 10	10 19	..	12 8	1 0	..	..	..
—	6 Denholme	6 46	7 26	7 58	..	..	9 13	10 22	..	12 11	1 3	..	..	..
—	7½ Thornton	6 50	7 30	8 2	..	..	9 17	10 26	..	12 15	1 7	..	..	..
—	8¾ Queensbury arr	6 53	7 33	8 5	..	..	9 20	10 29	..	12 18	1 10	..	..	..
—	**Queensbury** dep	6 59	7 34	8 10	..	8 37	9 25	10 32	..	12 22	1 16	..	1 35	..
6	Clayton	7 2	7 37	8 13	..	8 40	9 28	10 35	..	12 25	1 19	..	1 38	..
7¼	Great Horton	7 5	7 40	8 16	..	8 43	..	10 38	..	12 28	1 22	..	1 41	..
7¾	Horton Park	7 7	7 42	8 18	..	8 45	..	10 40	..	12 30	1 24	..	1 43	..
9	St. Dunstan's	7 11	7 46	8 22	..	8 49	..	10 44	..	12 34	1 28	..	1 47	..
9½	**Bradford** (Exchange).. arr	7 14	7 49	8 25	..	8 52	9 37	10 47	..	12 37	1 31	..	1 50	..
204¾	1 London (King'sCross) arr	12R12	..	..	..	1R45	2R45	5HR18	..	7SR15	..	..	..	..

Miles		p.m	p.m		p.m	p.m	p.m		p.m	p.m	p.m		p.m		Sundays	p.m	
		B	B		B	A	A		B	B	A		A			A	
	Halifax (Old) dep	..	3N35	..	4N41	5 10	5 43	..	..	..	7 50	..	10 25	..	..	4 18	..
¾	" (North Bridge)	..	3 38	..	4 44	5 13	5 46	..	..	..	7 53	..	10 28	..	..	..	..
2	Ovenden	..	3 42	..	4 48	5 17	5 50	..	..	..	7 57	..	10 32	..	..	..	..
2¾	Holmfield	..	3 45	..	4 51	5 20	5 53	..	..	..	8 0	..	10 35	..	..	..	..
5	**Queensbury** arr	..	3 51	..	4 57	5 26	5 59	..	..	..	8 6	..	10 41	..	..	..	..
—	Mls **Keighley** dep	1 41	3 18	..	4 28	4N55	..	..	5 46	6 31	7N27	..	10N15	..	..	..	..
—	1 Ingrow	1 47	3 24	..	4 34	5 1	..	..	5 52	6 37	7 33	..	10 21	..	..	..	..
—	3¾ Cullingworth	1 55	3 32	..	4 42	5 9	..	..	6 1	6 45	7 41	..	10 29	..	..	..	..
—	4¾ Wilsden	1 59	3 36	..	4 46	5 13	..	..	6 5	6 49	7 45	..	10 34	..	..	..	..
—	6 Denholme	2 2	3 39	..	4 49	5 16	..	..	6 8	6 52	7 48	..	10 37	..	..	..	..
—	7½ Thornton	2 6	3 43	..	4 53	5 20	..	..	6 12	6 56	7 52	..	10 41	..	..	..	..
—	8¾ Queensbury arr	2 9	3 46	..	4 56	5 23	..	..	6 15	6 59	7 55	..	10 44	..	..	..	..
—	**Queensbury** dep	2 11	3 56	..	5 3	5 27	6 0	..	6 17	7 7	8 9	..	10 50	..	..	..	..
6	Clayton	2 14	3 59	..	5 7	5 30	6 3	..	6 20	7 10	8 12	..	10 53	..	..	..	..
7¼	Great Horton	2 17	4 2	..	5 10	5 33	6 6	..	6 23	7 13	8 15	..	10 56	..	..	..	..
7¾	Horton Park	2 19	4 5	..	5 12	5 35	6 8	..	6 25	7 15	8 17	..	..	..	..	..	..
9	St. Dunstan's	2 23	4 9	..	5 16	5 39	6 12	..	6 30	7 19	8 21	..	..	..	..	..	..
9½	**Bradford** (Exchange).. arr	2 26	4 12	..	5 19	5 42	6 15	..	6 35	7 22	8 24	..	11 4	..	..	4 40	..
204¾	1 London (King'sCross) arr	7R46	..	..	..	..	..	..	2a49	..	..	..	..	..	..	9 30	..

A Through Train between Bradford and Halifax. a a.m. B Through Train between Bradford and Keighley. E Except Sats.
H Via Holbeck. Arr. 4 35 p.m on Fridays and Saturdays. J Except Saturdays. Dep. 10 55 p.m. on Fridays and 11 0 p.m. on Sundays.
N Through Trains between Halifax and Keighley. p p.m. R Restaurant Car. S Saturdays only. T Dep. 6-10 p.m. on Fridays.
Y Arr. 3 mins. *earlier* Z Arr. 4 mins. *earlier*

Table 94

L·N·E·R

Above: **Although it** is 6 September 1953, more than eighteen months before the line lost its passenger service, this view of Clayton station on the occasion of a joint SLS/MLS rail tour already indicates a railway in decline. The photograph, taken from the west, shows the signalbox in the foreground with the station's island platform beyond. The tour, which was hauled by Class N1 0-6-2T No 69430 – the prototype of the class introduced in 1907 – traversed many of the ex-GNR lines in the district, including the section from Shipley to Laisterdyke as well as the ex-MR branch from Guiseley to Yeadon. Freight traffic continued to be handled at Clayton until 10 April 1961 although the line remained operational for a further four years to serve the yard at Thornton. Following final closure, the site at Clayton was cleared and has been redeveloped as part of a housing estate. *Neville Stead/Transport Treasury*

Opposite above: **On** 5 June 1954, No 69474 is seen again on the line from Queensbury to Keighley, this time with a southbound service towards Queensbury. The station at Wilsden, which was some two miles from the place it purported to serve, was opened on 1 July 1886. The line from Thornton to Keighley was opened in various stages during 1884; it was not until 1 November of that year that passenger services reached Keighley. Situated south of Hewenden Viaduct, the station possessed a goods yard with freight traffic surviving until 11 November 1963 and the final closure of the section of the line north from Thornton. In 1908 a private siding had been added to serve a local quarry. Although the station itself was demolished, the goods shed at Wilsden remains intact and the trackbed north from Wilsden to Cullingworth over the viaducts at Hewenden and Cullingworth was converted into the Great Northern Trail in 2005. *Neville Stead Collection/Transport Treasury*

Opposite below: **In 1955,** shortly before the withdrawal of passenger services, Class N2/2 No 69541 – one of the batch with the condensing equipment removed following transfer from the London area – stands in Great Horton station with a Bradford-bound two-coach train. Although a relatively busy passenger station, it was the extensive freight yard at Great Horton that was an important source of traffic for the route. The final section of the Queensbury Triangle lines to remain operational was that from St Dunstan's to Great Horton, which survived through until 26 August 1972 and the final withdrawal of freight facilities from Great Horton. With redevelopment of the site, there is now virtually no indication that a railway once existed at this location. *Neville Stead Collection/Transport Treasury*

BRADFORD AREA • 29

Above: **Although promoted** by two smaller companies – the Bradford, Eccleshill & Idle and the Idle & Shipley railways – the line from Laisterdyke to Shipley was largely funded by the GNR and was operated from opening by that company. The single-track branch extended for 6¼ miles and served a second station in Shipley. From late 1875 there was a physical connection between the GNR and MR at Shipley. Like a number of railway lines, the circuitous line via Idle and Thackley suffered from the development of the local tramway network in the early twentieth century and passenger services were withdrawn on 2 February 1931. The route, however, remained operational by freight traffic until the 1960s. Here Class J39 0-6-0 No 64872 is seen heading southbound, having just passed through the closed Thackley station. The route was severed in November 1964 when the section from the erstwhile English Electric Phoenix Works at Thornbury to Idle was closed. The section from Idle to Shipley survived for stone traffic from a quarry at Idle until October 1968. The rump of the route, to the Phoenix Works, survived until August 1979. *Neville Stead Collection/Transport Treasury*

Opposite above: **In 1959,** Ivatt-designed 2-6-0 No 43070, which was allocated to Manningham briefly between January and November that year, is pictured on the slow lines at Frizinghall with a service destined for Valley Road goods. The first railway to serve Bradford – the Leeds & Bradford – opted for the easiest, rather than the most direct, route between Leeds and Bradford; this meant following the Aire valley as far as Shipley and then the Bradford Beck through Frizinghall and Manningham. In so doing, the railway replicated the route followed by the earlier Bradford Canal, which terminated in a basin close to the new railway station. The choice of route resulted in the Leeds & Bradford route extending for 13½ miles as opposed to the 10½ of the later line from Leeds Central to Bradford Exchange. *D. Butterfield/Neville Stead Collection/Transport Treasury*

Opposite below: **With Manningham** shed visible on the extreme right, Fairburn 2-6-4T No 42139 enters the station with a Forster Square-bound service in 1959. The MR opened its first shed at Manningham in 1872; the original section comprised a roundhouse, but this was supplemented by a further four-road shed to the south in 1887. The later addition was demolished by the LMS just before nationalisation with BR closing the roundhouse on 30 April 1967; the building was subsequently demolished. *Neville Stead Collection/Transport Treasury*

BRADFORD AREA • 31

Above: **Having been** held by a signal check, two Thompson-designed Class B1 4-6-0s – Nos 61110 and 61013 *Topi* – are pictured departing from Gildersome West with the 12.40pm service from King's Cross to Bradford Exchange on 6 August 1960. The 10¼-mile line from Laisterdyke to Ardsley was promoted by the Leeds, Bradford & Halifax Junction Railway with passenger services commencing on the section from Laisterdyke to Gildersome on 20 August 1856; freight traffic commenced as far as Gildersome on 1 January 1857 with the route being opened through to Ardsley on 10 October 1857. Although the intermediate stations along the route had all closed by the end of 1961, through services continued to use the line until 4 July 1966. The route was severed on 31 October 1966 when the section between Gildersome West and Birkenshaw & Tong was closed completely. *Mike Mitchell/Transport Treasury*

Opposite above: **Following the** conversion of the Bradford Moor trolleybus service to motorbus operation on 17 November 1962, trolleybuses on the routes along Manningham Lane made use of a turning circle in front of Forster Square until these services were converted on 31 October 1963. Pictured as passengers board during this period is No 755, one of a batch of eight delivered during 1950 and 1951 that represented the corporation's last wholly new trolleybuses. The background is dominated by the work of Charles Trubshaw, the MR's architect who oversaw the redesign of the station in 1890. On the left is the Midland Hotel with the arcaded wall that once provided access to and from the glass roofed carriage drive that fronted the station concourse. At the opposite end of the arcade are the goods offices for the Valley Road goods yard. Only the latter has been demolished as has much of the arcade, the section of the latter adjacent to the hotel and the hotel itself remain and are both now listed Grade II. *Author's Collection*

Opposite below: **There is** an increasing air of decay about Gildersome West station – witness the missing slates off the roof – on 16 July 1963 as Class J50 0-6-0T No 68935 heads west with a freight towards Bradford. The station, which was known simply as Gildersome when it first opened in 1856, gained the suffix 'West' on 2 March 1951, four years prior to its closure to passenger services on 13 June 1955. Although the section west from Gildersome to Birkenshaw & Tong closed completely on 31 October 1966, freight was handled on the section from Morley to Gildersome West until 16 March 1968. More than half a century after the line's final closure, this location is unrecognisable, with the area now dominated by the roundabout between the A650 and A62 and the connection to the M62 and M621 interchange. *Mike Mitchell/Transport Treasury*

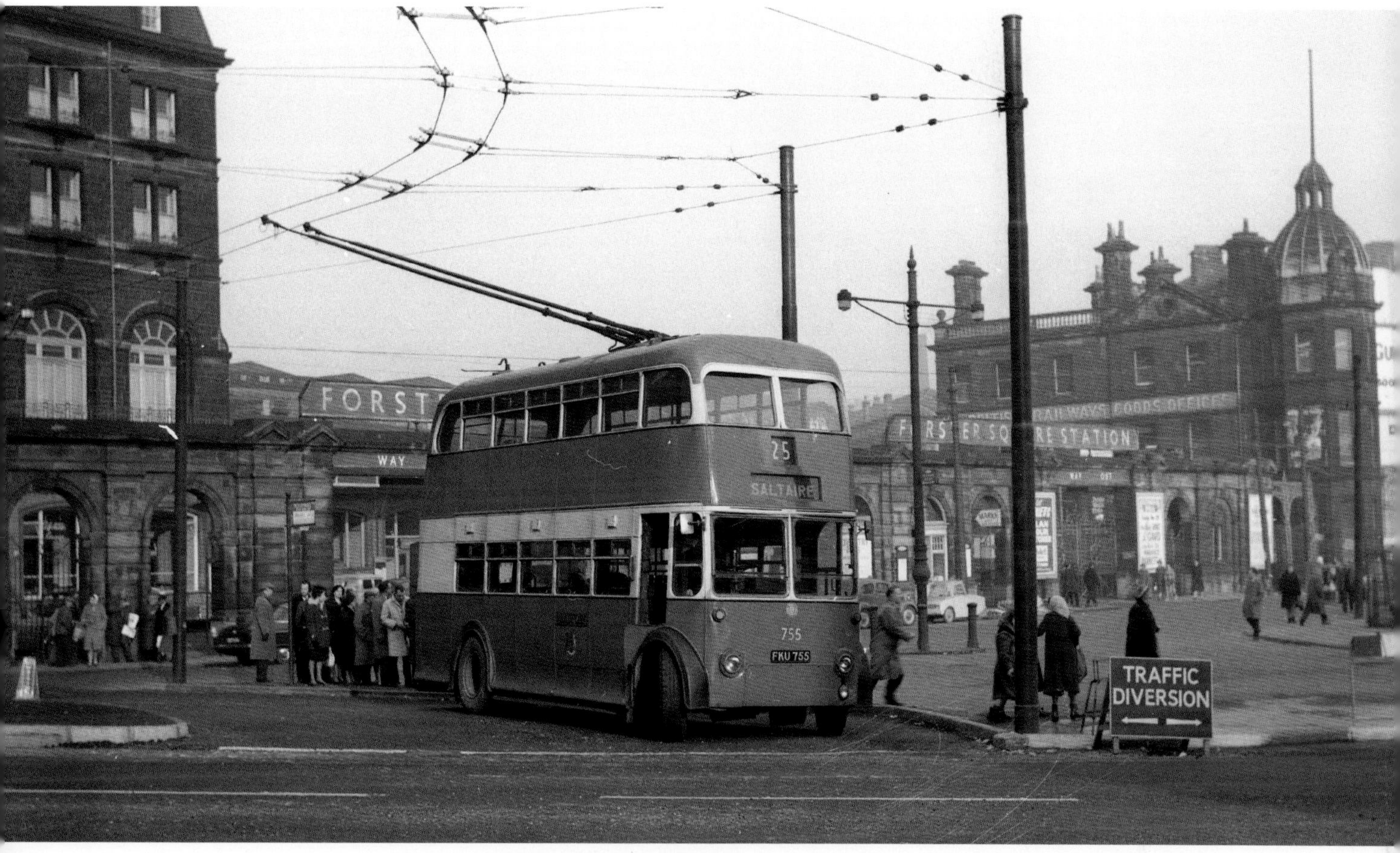
25
SALTAIRE
755
FKU 755
TRAFFIC
DIVERSION

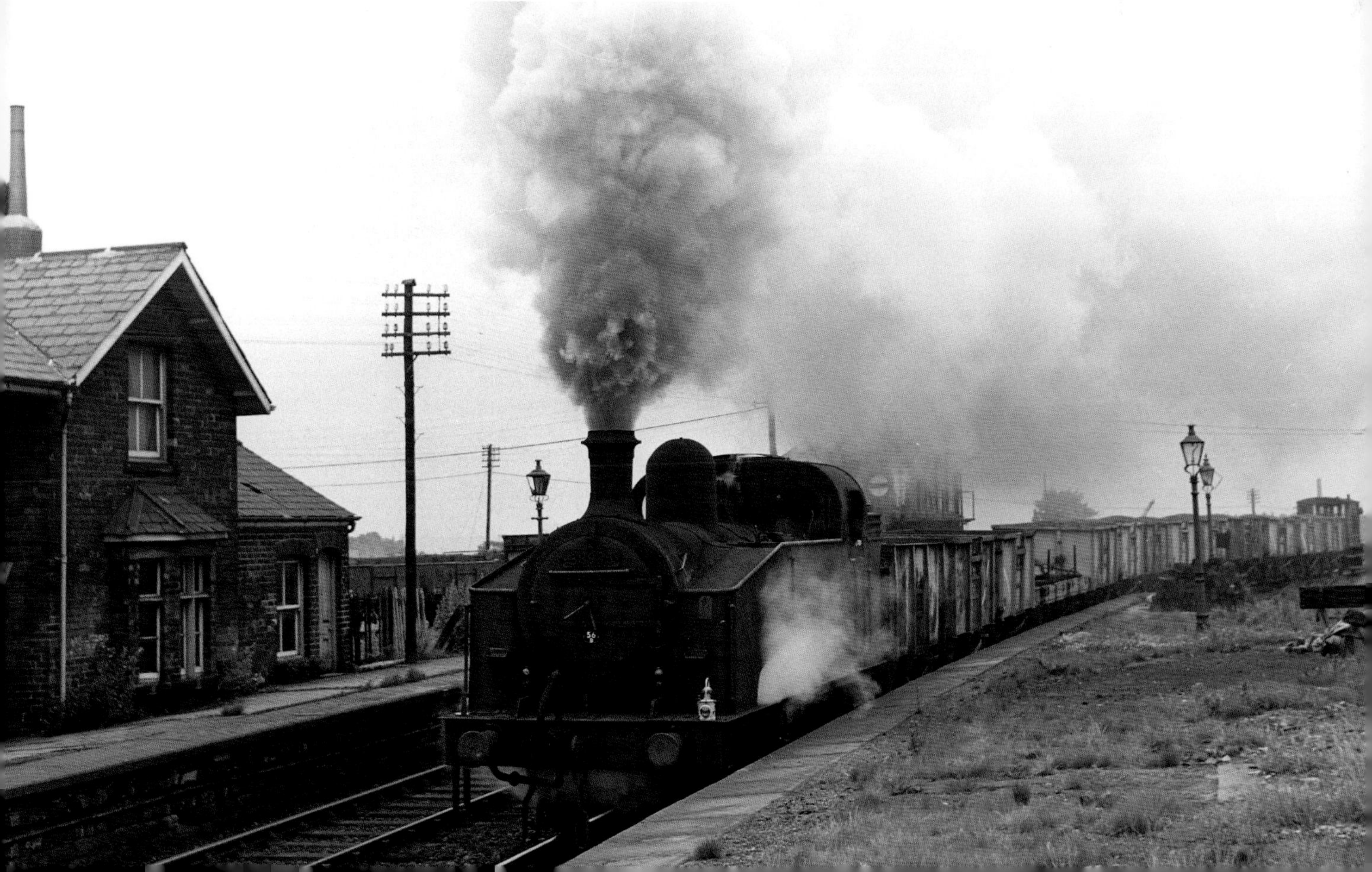

Above: **On 28 July** 1964, 2-6-4T No 42406 ascends the gradient from Bradford Exchange towards Laisterdyke with the 3.9pm service to King's Cross viewed from the Planetrees Road overbridge. The scale of the railway infrastructure at this point is impressive. The northernmost pair of lines are those that linked Laisterdyke with Adolphus Street, whilst the pair heading past the carriage sidings in the background are the avoiding lines towards Bowling Junction. The final pair of lines provided access to and from Planetrees Yard. The line from Leeds through to Bowling Junction along with the branch to Bradford Adolphus Street both opened on 1 August 1854; it was not until 7 January 1867 that the connection from Hammerton Street through to Mill Lane Junction was opened; this permitted GNR services to reach Exchange station. The view today is radically different with, effectively, only the two running lines from Bradford Interchange towards Leeds still regularly operational. The lines through to Adolphus Street closed when freight traffic to the terminal ceased on 1 May 1972. Following the withdrawal of passenger traffic over the line towards Bowling Junction on 6 January 1969, the avoiding line was effectively converted into a long siding for access to and from the steel terminal at Dudley Hill. After this traffic ceased, it remained used for traffic from the scrapyard established on the site of Planetrees Yard until 1985 when a new connection was installed at Laisterdyke; this permitted the final closure of the section through to Bowling Junction. Traffic from the scrapyard has now ceased. *Mike Mitchell/Transport Treasury*

Opposite above: **On 3 August** 1964, Class B1 4-6-0 No 61034 *Chiru* is pictured passing Cutlers Junction signalbox at Laisterdyke with the 3.9pm service from Bradford Exchange to King's Cross. The train is taking the line towards Dudley Hill; the lines heading off to the east at this point formed the alternative route to Leeds via Pudsey Greenside and Pudsey Lowtown. Bu this date, however, the Pudsey loop had closed. Passenger services had ceased on 15 June 1964 and freight traffic followed on 6 July 1964. The section of line from Laisterdyke to Dudley Hill was the final section of the once through route to London to survive; having lost its passenger services on 4 July 1966, the route was closed to freight in stages. The section south from Dudley Hill to Birkenshaw & Tong closed on 16 March 1968 with the final section to Dudley Hill succumbing in August 1979. Although there has been some landscaping at this location over the past 40 years, it is still possible to identify the location of the disused trackbeds at this location. *Mike Mitchell/Transport Treasury*

BRADFORD AREA • 35

Left: **Redevelopment in** the city centre allied to the demolition of many of the Victorian buildings in and around Forster Square resulted in the exterior of Exchange station becoming much more visible. With evidence of demolition in the foreground and with the trolleybus overhead along the southern section of Hall Ings, this was the view of the station's exterior on 6 September 1964. Although the bulk of the station was demolished following its closure, the lie of the land meant that the lower level of the building – including the entrance seen here – survived; the latter is still extant and provides a route up from the A6181 to the Crown Court, a building that was constructed on the site of the closed station. The original LYR station in Bradford – Drake Street – opened on 9 May 1850; this was a relatively small affair that comprised a single island platform. With the arrival of the GNR in 1867, a second island platform was added, giving a total of four platform faces. However, as the local network and the number of trains grew, so the original station became increasingly inadequate. As a result, a new ten-platform station, with a two-span overall roof designed by William Hunt (an LYR engineer), was completed in 1888; the station was wholly owned by the LYR with that company using platforms 1 to 5 with 6 to 10 being allocated to GNR services. *R.B. Parr/John Meredith Collection/Online Transport Archive*

Above: **Devoid of** its smokebox numberplate, Standard Class 2MT 2-6-0 No 78014 stands at the buffer stop of platform 6 at Bradford Forster Square during 1964. The appearance of this class of locomotive in the Bradford and Leeds area was relatively uncommon; indeed, only one example – this particular locomotive – was allocated to any shed in the area and then only briefly. No 78014 was based at Farnley Junction from November 1963 until April 1964 when it was transferred to Manningham; its sojourn in the West Riding ended in December 1964 when it was transferred across the Pennines to Gorton in Manchester; it was withdrawn in September the following year having achieved just over eleven years in service. The station as illustrated here was the result of the MR work completed in 1890. The original Leeds & Bradford Railway station had opened on 1 July 1846; this was rebuilt by the MR in 1853 but the 1890 reconstruction included six platforms, a ridge and furrow overall roof, hotels and goods station. However, the ridge and furrow roof was removed in the early 1950s to be replaced by platform canopies. *Neville Stead Collection/Transport Treasury*

Opposite above: **It must** be about 7.15am as two Class B1s – Nos 61014 *Oribi* and 61240 – head on to the line towards Dudley Hill at Laisterdyke East Junction with the 7.8am Saturdays Only service from Bradford Exchange to Skegness on 1 August 1964. At this date No 61014 was a relatively recent arrival in the area; it had been transferred from Tyne Dock to Low Moor on 7 June 1964. It would remain based there until a final reallocation took it back to the north-east – at Blyth (North) – on 21 August 1964. No 61240 was a longer resident in the West Riding; based at Ardsley since November 1962, it would end its days at Wakefield from where it was withdrawn in December 1966. *Mike Mitchell/Transport Treasury*

Left: **Accidents** have always been a sad feature of railway life; one of the more spectacular in the Bradford area occurred on 10 November 1964 when Ivatt-designed Class 4MT No 43072, working a twenty-one-wagon empty wagon train from Ardsley to Bradford Adolphus Street, ran out of control on the approach to the station running through the yard. Fortunately, the crew managed to leap from the cab before the locomotive broke through the buffers and ended up in the mess seen here in Dryden Street. The locomotive was cut up on site four days later. Bradford Adolphus Street was the original GNR terminus in the city. It opened on 1 August 1854 with a single train shed spanning the six tracks and four platforms. The station was, however, inconvenient – there was a considerable walk from the centre up a steep hill – and services ceased on 7 January 1867 when the line through to Drake Street (later Exchange) was opened. Adolphus Street continued to handle freight traffic until final closure on 1 May 1972. Although the station was demolished to facilitate the widening of Wakefield Road, it is still possible to see traces of the structure – including a bricked-up entrance – on Dryden Street. *Transport Treasury*

Above: **On 6 September** 1965, the RCTS (West Riding Branch) organised the 'West Riding' rail tour, using a DMU to traverse a number of freight-only lines in the area. The train was booked to start from Bradford Exchange at 10.25am and the first line to be covered was the section of the ex-GNR 'Queensbury Triangle' lines from St Dunstan's to Thornton via Queensbury. These routes had lost their passenger services in May 1955 and had closed completely between Thornton and Denholme in November 1963. Thornton had once seen well-tended platforms and gardens; this was the scene of destruction witnessed by the passengers on the special. Less than year later – in June 1965 – the line from Horton Park to Thornton closed completely. Other routes visited by the special included the City Road and Kirkburton branches. *R.B. Parr/John Meredith Collection/Online Transport Archive*

Opposite above: **In 1965**, rebuilt 'Royal Scot' No 46115 *Scots Guardsman*, then allocated to Carlisle (Kingmoor), awaits departure from Bradford Forster Square. Completed by the North British Locomotive Co in Glasgow during October 1927 and rebuilt twenty years later, No 46115 was preserved following withdrawal in January 1966. Following preservation, the locomotive was based on the Keighley & Worth Valley Railway but, deemed at the time to be unsuitable for use on the branch, it moved in May 1969 to the Dinting Railway Centre. After restoration it made two main line runs during 1978 but needed a major boiler overhaul. Transferred from Dinting to Tyseley in 1989, its restoration to main-line condition was finally completed at Carnforth, courtesy of owners West Coast Railway Co, in 2008 and it remains main-line certified. *Neville Stead Collection/Transport Treasury*

Opposite below: **With Laisterdyke** station – closed 4 July 1966 – in the background and Quarry Gap Sidings in the foreground, Class B1 4-6-0 No 61161 has just passed Laisterdyke East signalbox on 3 July 1965 to take the route to Leeds via Stanningley with the 8.30am (Saturdays Only) service from Bradford Exchange to Cleethorpes via Leeds. *Mike Mitchell/Transport Treasury*

Above: **On** 5 July 1965, 2-6-T No 42055 has just departed from Low Moor with a service heading towards Halifax. The background is dominated by the extensive carriage sidings that were once a feature of the triangle at Low Moor. The lines heading off to the right form the southern part of the triangle and connected into the line that ran south from Low Moor towards Cleckheaton. This curve was taken out of use finally on 1 July 1968. The original line to serve Low Moor was the LYR route from Mirfield; this opened on 18 July 1848 and was extended through to Bradford Drake Street on 9 May 1850. The line from Halifax to Low Moor followed on 7 August 1850. *Mike Mitchell/Transport Treasury*

Opposite above: **Seen entering** the closed station at Dudley Hill in 1966 with a southbound service is 2-6-4T No 42184. Although the station had closed some fourteen years earlier – on 7 April 1952 – it still looks in reasonable condition with a platform seat still visible on the Up platform. This was the second station to serve Dudley Hill; the original station, situated slightly to the south, had opened on 20 August 1856 and had been relocated on 1 October 1875 in connection with the line to Low Moor. The freight yard, visible in the background, was to remain open for more than a decade after the withdrawal of passenger services over the line; freight facilities were officially withdrawn from Dudley Hill in December 1979. *Neville Stead Collection/Transport Treasury*

Opposite below: **With the** looming bulk of the former corporation power station in the background, 'Jubilee' class No 45675 *Hardy* is pictured departing from Valley Road during 1967 with a mixed freight. Valley Road was the MR's primary freight yard serving Bradford and the facilities provided included a bonded warehouse. *Neville Stead Collection/Transport Treasury*

BRADFORD AREA • 41

Above: **By June** 1967, steam operation in the Bradford area was drawing to a close and, on the 10th of that month, Fairburn-designed 2-6-4T No 42252 is seen ascending the gradient from Exchange station towards Mill Lane Junction. The locomotive, which had been completed at Derby works in November 1946, was a relatively recent transfer to the Eastern Region, having been allocated to Tebay for a brief period from 31 December 1966. It moved to Eastern Region at the end of May 1967, being officially allocated to Low Moor on 17 June 1967. A final transfer took it to Normanton the following month, from where it was withdrawn at the end of September 1967. *Alan Murray-Rust/Online Transport Archive*

Opposite above: **On the** same day, another of the Fairburn 2-6-4Ts – No 42145 – is pictured taking water at Exchange. New in June 1950, No 42145 had spent twelve years allocated to Carstairs, on Scottish Region, prior to a transfer to Ardsley October 1963; less than a year later, in September 1964, the locomotive moved to Holbeck from where it was withdrawn in August 1967. Following withdrawal, No 42145 spent some time in store at Low Moor before being scrapped in February 1968. *Alan Murray Rust/Online Transport Archive*

Opposite below: **Although timetabled** main-line steam ceased in August 1968, one locomotive – the preserved Class A3 4472 *Flying Scotsman* – continued to appear as the locomotive's then owner, Alan Pegler, had a contract with British Rail that permitted the running of steam-operated specials through until 1972. Following the locomotive's overhaul during the winter of 1968/69, No 4472 undertook a number of excursions during 1969 before it crossed the Atlantic for its ill-fated trip to North America. On 1 June 1969, Flying Scotsman Enterprises organised the 'Grand Tour of the North'; this train operated from Doncaster via Leeds, Manchester, Preston, Newcastle and Leeds before returning to Doncaster. The section from Leeds to Manchester brought No 4472 with its train to the Laisterdyke to Bowling Junction line and the train is pictured here approaching the Hall Lane level crossing. The Bradford avoiding line was officially closed to passenger traffic on the same day. The route from Bowling Junction survived – effectively as a long siding – until 1985 to serve a scrapyard at Laisterdyke; with the track being reconnected at Laisterdyke to serve the yard, the section from there to Bowling Junction was closed and lifted. *Michael H. Waller/Author's Collection*

BRADFORD AREA • 43

The approach to Bradford Exchange viewed from the west on 21 March 1972. Prominent to the right of the actual entrance is the cabmen's shelter. Erected in 1877, the shelter provided accommodation for the use of the drivers of horse cabs and, following closure of the station, was donated by the members of the Bradford Stations Taxi Association in January 1973 to the tramway museum at Crich in Derbyshire. At the time of writing, the National Tramway Museum had just completed the restoration of the structure to its original 1877 form, including clerestory roof, with work being supported by the Pilgrim Trust, the Arts Council National Lottery Project and the Association for Industrial Archaeology. When the structure was originally completed, *The Building News* commented on one of its features: 'The stove, which is very compact, contains oven, hot-plate, and boiler for supplying warm water for the horses – an arrangement much appreciated by the cabmen.' *John Meredith/Online Transport Archive*

Viewed from the south on 21 March 1972, Exchange station was already much reduced; the track to platforms 1 to 5 had been removed as part of the preliminary work for the construction of the initial two platforms for the replacement station. The Bridge Street bridge, the weakness of which was one of the major contributory factors to the demise of the original station, is visible in the foreground; by the date of the photograph, a two-ton weight limit was in force over the bridge. Pictured entering the station is a three-car Class 110 'Calder Valley' DMU with No E51843 bringing up the rear. The thirty-strong Class 110 'Calder Valley' units were synonymous with services into Bradford Exchange for many years; ordered in March 1959 and built during 1961 and 1962 by the Birmingham Railway Carriage & Wagon Co, the units were fitted with powerful 180hp Rolls-Royce engines in order to cater for the gradients encountered on the Leeds to Manchester via Bradford route. Initially twenty sets were allocated to Hammerton Street and the remainder to Newton Heath but, by the mid-1970s, all the surviving units were based at the former (one three-car set was scrapped in November 1963 and another power car in 1972). Following refurbishment during 1979 and 1980, a number of the centre trailers were withdrawn in the early 1980s; these were subsequently replaced by trailers from withdrawn Class 111 units. As a result of the closure of Hammerton Street in 1984, the surviving units were transferred to Neville Hill in Leeds. All were withdrawn by the end of 1991 with one two-car and one three-car set being preserved. *John Meredith/Online Transport Archive*

The twilight of Bradford Exchange: although the track in platforms 1 to 5 remains intact, it has been disconnected south of Bridge Street with the surviving services restricted to platforms 6 to 10. The glazing has been removed from the southern section of the impressive train shed. Notice that much of the steam era infrastructure – including the water tower and column (at the end of platforms 7 and 8) – is still extant whilst an '03' shunter is stabled in the carriage sidings to the east. *John Meredith/Online Transport Archive*

The concourse at Exchange on 21 March 1972; for almost a century this was a familiar sight to travellers to and from the city. Passengers mingled with trolleys carrying the mail whilst, on the extreme right, another familiar sight was the W.H. Smith kiosk is visible. The latter played an important role for me; it was where, as a schoolboy, I purchased my first Ian Allan publications; little did I know then that I would end up working for the publisher for more than 25 years! *John Meredith/Online Transport Archive*

With the looming bulk of Valley Road power station in the background, two grounded ex-Pullman coaches were situated on the approaches to Bradford Forster Square. The concept of the luxury train was pioneered by the Chicago-based George Pullman and was first introduced to Britain, in conjunction with Pullman, by the Midland Railway in June 1874 on a service from London St Pancras to Forster Square. The original carriages for the MR's service were constructed from parts imported from the Pullman Palace Car Co's workshops at Detroit in the USA and, with their balconies at both ends, were very American looking in appearance. The two grounded bodies illustrated here were amongst the original eight supplied for the new service and survived in service until all were withdrawn between 1888 and 1900. Four of the eight – including these two – were transferred to departmental use. The two bodies were salvaged in 1975 and transferred to the Midland Railway Project at Butterley. It is believed that only one now survives and that is in a poor and unrestored condition. *John Meredith/Online Transport Archive*

With the two grounded ex-Pullman coaches of 1874 on the extreme left, this view from the yard at Valley Road shows the considerable length of footbridge 7A as it crosses the railway at this point. Visible again is Valley Road power station. Following the passage of the Electric Lighting Act in 1882, individuals, companies and local authorities were empowered to construct and operate power stations to generate electricity. Bradford Corporation was the first municipality to operate a power station; this first station was opened in 1887 on Bolton Road. By the middle of the next decade, this station was inadequate for the needs of a growing city, and, in June 1896, work commenced on the construction of a new coal-fired power station on Valley Road. In order to minimise pollution – already a significant problem in the valley – the new station's chimney had to be a minimum of 90ft in height. At its peak, the power station consumed some 200 tonnes of coal per day; this was transported by rail to Valley Road and transferred to the power station by conveyor. Nationalised in 1948, Valley Road power station finally closed in October 1976 and the site was later cleared for redevelopment. *John Meredith/Online Transport Archive*

Above: **The depressing** scene at Bradford Exchange on 21 July 1973 as work on the demolition of the old terminus has seen much of the trainshed removed. Although the station had closed earlier that year, it was not until much later in the decade – following the completion of the new interchange – that the last sections of the old station were finally eliminated, with work continuing through until early 1982. The site of the station was subsequently used for the location of the city's new Crown Court. *Chris Gammell/Online Transport Archive*

Opposite above: **The twenty-two-strong** Class 55 'Deltics' were a familiar sight on express services from Bradford and Leeds to King's Cross for almost two decades until they were supplanted by HSTs. Despite their power, however, they often struggled to depart from Bradford Exchange in adverse weather conditions. With the derelict remains of the old station in the background, No 55019 *Royal Highland Fusilier* is pictured here departing from the new station with the 11.55 service to King's Cross. Withdrawn at the end of December 1981, No 55019 is one of six of the type to survive in preservation. *Gavin Morrison*

Opposite below: **The closure** of the original Exchange and the station's relocation south of Bridge Street was part of a larger scheme to provide an integrated transport interchange comprising the railway station, large bus station and underground bus depot. The bus station and depot were built on the site of the former LYR Bridge Street goods shed. Although the first two platforms of the new railway station opened in 1973, it was to be four years before the project was completed. The opening of Bradford Interchange took place on 20 March 1977 and saw the first visit of an HST to the city when No 253021 formed a special. In the foreground Class 47 No 47409 is seen awaiting departure with a service for King's Cross. The railway station was still known as Bradford Exchange at the time; it was not until 16 May 1983 that it became officially renamed Bradford Interchange. Changes in policy – most notably privatisation of the local bus companies and deregulation of bus services – in the late 1980s rendered the bus station increasingly underutilised, with the result that its facilities have been much reduced. *Gavin Morrison*

BRADFORD AREA • 51

A familiar sight for many years at Bradford Exchange/Interchange was the stabled Class 03 shunter with match wagon parked on the siding to handle – as and when required – the parcels stock. Hammerton Street had an allocation of these shunters for many years and on 5 July 1977, No 03371 is in residence at the station. How much work these locomotives actually did was – and remains – a mystery to me as, travelling through the station on numerous occasions, I never actually saw an '03' moving! New in December 1958 as departmental No 92 and allocated to Lowestoft, the locomotive, having been renumbered D2371 in August 1967, was first transferred to Hammerton Street in September 1968. Based at Healey Mills and York (North) between March 1973 and July 1976, when it returned to Hammerton Street, the locomotive was finally withdrawn in November 1987 and was subsequently preserved. *Author*

With the Bradford to Leeds line curving away in the background, Laisterdyke East signalbox is pictured here on 19 April 1980 towards the end of its existence. Work is already in progress in removing the track from the Quarry Gap Sidings. The surviving sections of the ex-GNR routes to Shipley and to Wakefield – to the Phoenix Yard siding that served the former English Electric factory at Thornbury and to the freight yard at Dudley Hill respectively – had both been officially closed on 6 August 1979. *Author*

Pictured in the platforms of the new Bradford Forster Square station in 2001 are two of the Class 333 EMUs for use on the Airedale and Wharfedale lines. Services ceased to use the original Forster Square station on Saturday 9 June 1990, with buses being used on the following day. The new three-platform station, constructed slightly to the west of the original, was brought into use on Monday 11 June 1990. The bulk of the remains of the original station were demolished two years later. The site of the old station and the adjacent Valley Road goods shed has been partially redeveloped. *Geoffrey Tribe/Online Transport Archive*

As passengers wait at the Up platform for a service towards Shipley on 30 June 2010, Class 333 No 333012 heads towards Bradford Forster Square. The original station at Frizinghall, closed on 22 March 1965, had two platforms opposite each other; when the station reopened on 7 September 1987 the station on the Down side was relocated to the south of Frizinghall Road. One of major sources of traffic to and from the station are the pupils that attend the adjacent Bradford Grammar School, and it was one of the teachers from the school, the late Dr Robin Sisson (with whom I walked through Summit Tunnel prior to its reopening following the disastrous fire in 1984), who was one of the leading campaigners for the station's reopening. *Gavin Morrison*

THE MIDLAND LINES FROM LEEDS

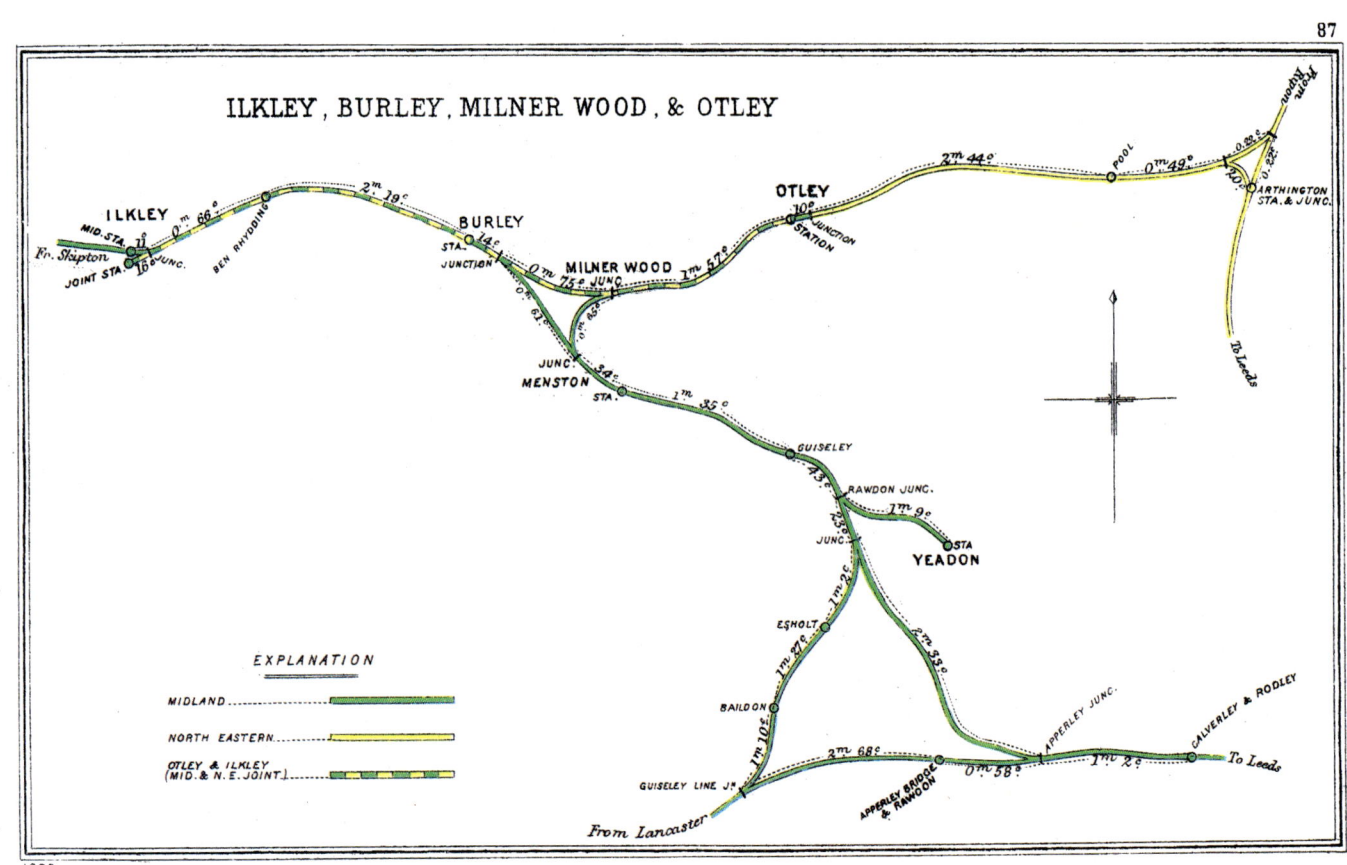

Above: **The network** of lines operated by the MR and NER in the Ilkley area as recorded in the Railway Clearing House map of 1905.

Opposite above: **The LMS** timetable for the period from 16 June to 5 October 1947 covering the lines from Bradford Forster Square and Leeds to Skipton via either Keighley or Ilkley.

Opposite below: **Although the** terminus of the 1¼-mile Yeadon branch was provided with a passenger station, there were never any timetabled passenger services on the line. However, the occasional special did traverse the route and, on 6 September 1953, the 'West Riding' rail tour, jointly organised by the MLS and SLS, visited the branch. The train, which was hauled by Class N1 0-6-2T No 69430, is pictured in the platform at Yeadon. The first proposal for a line serving Yeadon was in 1881 when plans for a link from Horsforth to Guiseley via Yeadon were proposed; the NER was unwilling to construct the line with the result that the MR-backed Guiseley, Yeadon & Rawdon Railway was authorised by an Act of 16 July 1885 to construct a line. However, funding was impossible and a new Act – of 5 August 1891 – was required. Under its new name – the Guiseley, Yeadon & Headingley Railway – the company was taken over by the MR on 28 June 1892 prior to its opening on 10 August 1894. The line was closed temporarily as a wartime economy measure during 1944; it closed finally on 7 August 1964. *Tony Wickens/Online Transport Archive*

THE MIDLAND LINES FROM LEEDS • 55

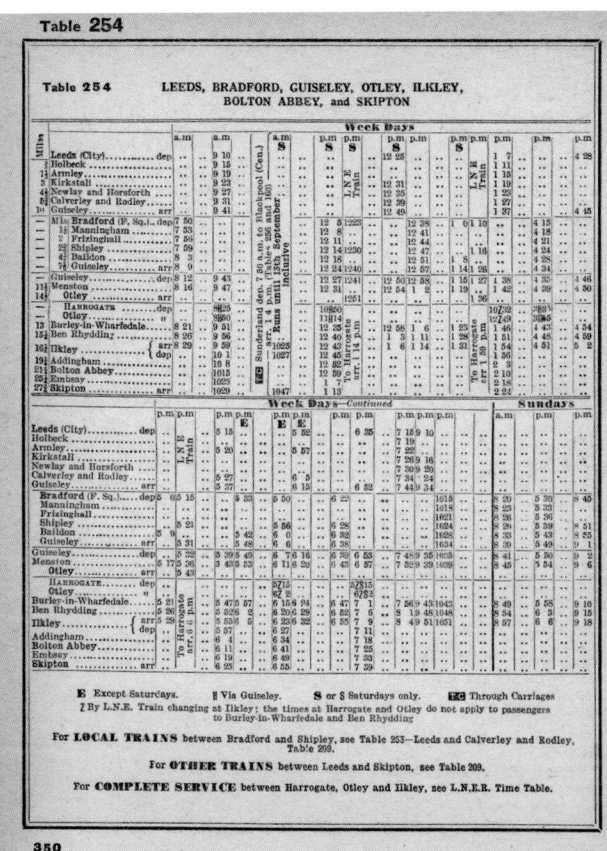

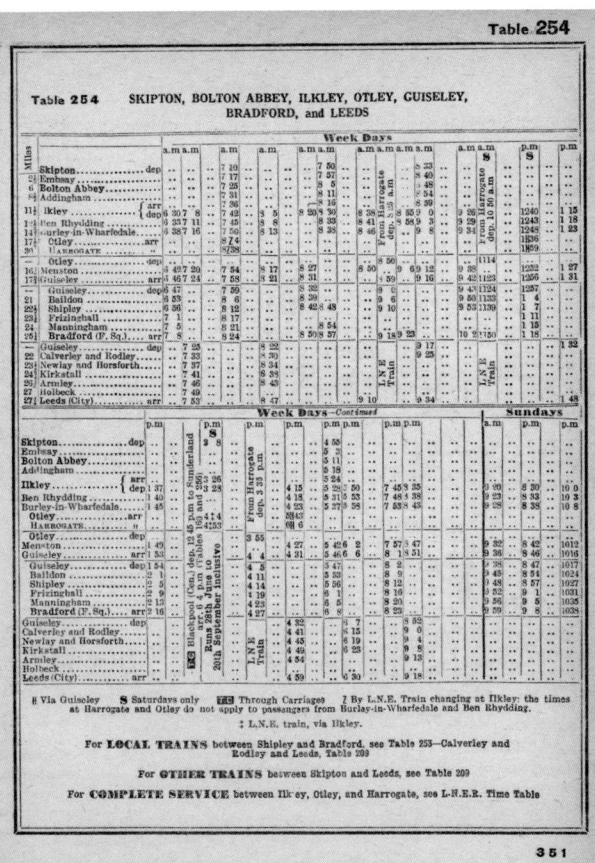

Above: **Grassington &** Threshfield station viewed from the north in 1955. The original Yorkshire Dales Railway had ambitious plans for the construction of a link from Skipton to Darlington; in the event, only the 8½ miles from Embsay Junction to Grassington was completed, with services commencing on 29 July 1902. Like a number of rural branch lines, the service to Grassington suffered from bus competition during the 1920s and the LMS withdrew the passenger service on 22 September 1930, although the branch remained a popular destination for walkers with a number of excursions operated to it. In terms of freight, agriculture was a mainstay, whilst a private siding, which headed to the west just south of the terminus, served the Threshfield limestone quarry of Settle Limes Ltd. *Neville Stead Collection/Transport Treasury*

Opposite above: **Viewed looking** towards Ilkley in 1958, Ben Rhydding station originally opened on 1 July 1866, some eleven months after the opening of the line to Ilkley. The original station was constructed in wood but, in May 1871, agreement was reached between the Otley & Ilkley Joint and the proprietor of the Ben Rhydding Hydro for the construction of a more impressive structure. This was to be built, subject to the design being approved by the railway's engineer, at the expense of the Hydro with the railway charging the peppercorn rent of 1d per annum for the land. In 1885 the railway acquired the buildings – the stone-built structure on the Down platform and a wooden shelter on the Up – for £240. The signalbox was constructed in 1901; this was to survive until closure in December 1965. Visible in the view is also the small goods yard that served the station; freight facilities were withdrawn on 5 July 1965. On 7 October 1968, following the introduction of 'Pay Trains' to the line, the station became unstaffed and the attractive buildings were subsequently demolished. *Neville Stead Collection/Transport Treasury*

Opposite below: **Stanier-designed** 2-6-2T No 40112 stands in Keighley platform No 1 with an eastbound service towards Shipley during 1958. The first station to serve the town was opened by the Leeds & Bradford Railway in March 1847 slightly to the west of the current location; the station was resited in 1883 in order to accommodate the arrival of services on the GNR route via Queensbury. Completed at Derby in July 1935, No 40112 had been allocated to Kentish Town in London before its transfer to Manningham during September 1953. It was to remain based there until a final move saw it reallocated to Copley Hill in early 1959 from where it was withdrawn in November 1962. *Neville Stead Collection/Transport Treasury*

THE MIDLAND LINES FROM LEEDS • 57

Above: **The Lancashire** branch of the RCTS organised the 'Roses Rail Tour' on 8 June 1958. The train, formed of two two-car DMUs (Nos M50806+M50773 and M50804+M50771), commenced its journey at Manchester Victoria and is pictured here in the Ilkley platforms at Skipton prior to heading to Harrogate, Wetherby and Church Fenton. *John McCann/Online Transport Archive*

Opposite above: **On 8 April** 1959, the Up 'Waverley' is pictured approaching Kirkstall station behind Holbeck-allocated 'Jubilee' No 45568 *Western Australia*. The 'Waverley' had its origins in the 'Thames-Forth' express that had been introduced in September 1927. Starting at London St Pancras, it ran via the Midland main line to Leeds and then onwards to Edinburgh via the Settle & Carlisle, Carlisle and the ex-North British Waverley route. Losing its name the start of the Second World War, the service was revived – as the 'Waverley' – in 1957 but taking nine hours fifteen minutes to complete the through journey meant that it suffered in comparison with other services (the 'Flying Scotsman', for example, took six hours to travel from King's Cross to Waverley via the East Coast main line). Poor time keeping was also a feature. The service became seasonal in 1965, operating during the summer months only, and was completely withdrawn in September 1968. *Peter D.T. Pescod/Transport Treasury*

Opposite below: **With Kirkstall** Power Station forming the backdrop, Class 4F 0-6-0 No 43871 heads an eastbound freight towards Leeds. The power station at Kirkstall was developed by Leeds Corporation following its purchase of the site in 1926; construction commenced two years later, with the first power being generated in October 1931. Originally coal fired, expansion of the station saw a considerable amount of coal carried to the site by rail and water, but this traffic was considerably reduced in 1964 when the power station was partially converted to oil firing. The power station was finally closed in 1976 and the building demolished; the cooling towers succumbed in 1979. *Neville Stead Collection/Transport Treasury*

THE MIDLAND LINES FROM LEEDS • 59

Above: **Viewed through** the rear cab windows of an eastbound DMU, the platforms at Kildwick & Crosshills are virtually empty on 21 August 1959. The first station here, situated adjacent to the level crossing with the A629, was opened by the Leeds & Bradford Extension Railway in 1847. It was originally known as Kildwick; it became Kildwick & Cross Hills in 1863 and Kildwick & Crosshills fifteen years later. A new station, as seen here, was opened on 7 April 1889. Ironically, although the later station closed on 22 March 1965 and was subsequently demolished, the station house of the original structure remains intact on the south side of the level crossing. *Henry Priestley/Transport Treasury*

Opposite above: **Viewed from** a Skipton-bound train on 21 August 1959, Addingham station looks empty. The station originally opened, with the extension of the line from Ilkley through to Skipton, on 16 May 1888. Passenger services were withdrawn west of Ilkley on 22 March 1965 with the station's freight facilities ceasing on 5 July 1965. Since closure, the bridge to the north of the station has been demolished and the station site itself has been redeveloped for housing. The Embsay & Bolton Abbey Steam Railway has a long-term aspiration to restore the section east from Bolton Abbey towards Addingham but, if achieved, will require the construction of a new station slightly short of the original site. *Henry Priestley/Transport Treasury*

The branch line service from Keighley to Oxenhope during the summer of 1947 – the last year of the LMS's independent existence. At this stage, Damems station was still open; it would be the first casualty on the line when passenger services were withdrawn in 1949. *Author's Collection*

Above: **The notionally** independent – but heavily backed by the MR – Keighley & Worth Valley Railway opened in 1867. The 4¾-mile long line's terminus was at Oxenhope, where Ivatt 2-6-2T No 41273 is pictured in 1960 having arrived with a three-coach train from Keighley. The line closed completely in June 1962 following the withdrawal of freight services but that was not to be the end of the story; following preservation, the line reopened six years and the crowds on the platform at Oxenhope when the line is in operation today are in stark contrast to the virtually empty visible more than six decades ago. *Neville Stead Collection/Transport Treasury*

Opposite above: **Class 9F 2-10-0** No 92167 takes the curve at Shipley from Bingley Junction to Leeds Junction with an Up freight during 1967. New In May 1958, No 92167 was, when recorded here, allocated to Saltley shed in Birmingham; like the majority of the class, the locomotive was destined to have only a relatively short operational career, being withdrawn from Carnforth in June 1968. Visible above the train is Shipley Bingley Junction signalbox. The box, which dated originally to 1907, was taken out of service in 1994 and, following purchase by the Leeds North West Signalling Preservation Group, was presented to the Keighley & Worth Valley Railway. After restoration, the box was commissioned to serve the preserved line's platforms at Keighley during 2018. *Neville Stead Collection/Transport Treasury*

Opposite below: **In 1961,** Fairburn-designed 2-6-4T No 42141, allocated at the time to Manningham, has just departed from Shipley with a service towards Bradford Forster Square. On the left can be seen the ex-MR goods yard that served Shipley; freight facilities were finally withdrawn from the yard in September 1980. Although the goods shed and much of track has now disappeared from the yard, the site is now occupied by the Crossley Evans scrap yard, which was first established in 1969. The yard was rail connected until the outgoing scrap was transferred to road transport in 2016. No 42141 was based in the West Riding for the bulk of its seventeen-year career; delivered new to Saltley in Birmingham during April 1950, it was transferred to Normanton in September 1952 and was based at Manningham from February 1959 until April 1964; it was finally withdrawn from Low Moor in late 1967. *Neville Stead Collection/Transport Treasury*

THE MIDLAND LINES FROM LEEDS • 63

Above: **On 10 May** 1961, Class 4F 0-6-0 No 43871 pulls out of the yard at Guiseley with a southbound freight service. Freight facilities were withdrawn from Guiseley on 7 October 1968, but coal traffic recommenced in December 1969; the run-down in coal traffic subsequently resulted in this traffic ceasing as well. Although the track had disappeared, the goods shed itself still survives in commercial use. The buildings on both platforms have been demolished; these were replaced by a simple shelter on the Down side. However, in 2002, more substantial facilities were constructed on both platforms. Following the closure of the signalbox, the structure was preserved and re-erected at Bolton Abbey. The MR line from Apperley Junction, on the Leeds to Shipley main line, to Menston Junction was authorised by an Act of 11 July 1861 and opened in conjunction with the Otley & Ilkley Joint line from Otley to passenger traffic on 1 August 1865 and to freight traffic on 1 October 1866. Guiseley became a junction with the opening of the line to Guiseley Junction, Shipley, to Esholt Junction to passenger traffic on 4 December 1876 and to freight traffic nine days later. *Henry Priestley/ Transport Treasury*

Opposite above: **In 1962,** Class 8F 2-8-0 No 48454 approaches Steeton & Silsden station with a Down freight service. The train is passing the station house, which is situated to the east of the level crossing, whilst the platforms were situated to the west (reflecting the relocation of the platforms in 1892). Also visible on the extreme left is the signalbox; this was an MR Type 4d design that was opened on 18 February 1923 to replace an earlier structure that dated to the 1870s. Known as Steeton Station from later in 1923, the box received a replacement frame in 1940. The station was closed on 22 March 1965 although the box which controlled the level crossing survived. The level crossing itself was abolished on 10 July 1898, following the opening of the new Skipton to Keighley road, and the box itself was closed on 29 October 1989. A new station, which effectively straddles the erstwhile level crossing, was opened on 14 May 1990. The 1940 Railway Executive Committee frame from the box was transferred to the North Norfolk Railway. The box itself was destroyed by fire on 17 June 1999 having been used for a number of years as a taxi office. *Neville Stead Collection/Transport Treasury*

Opposite below: **On 23 June** 1962, some six months after the line had lost its passenger services and a week after freight services were withdrawn, the Keighley & Worth Valley Preservation Society organised a special from Bradford Forster Square hauled by Manningham-allocated Class 3F 0-6-0 No 43586 to mark the line's final closure. Here the train is seen at Oakworth on its outbound journey. *Transport Treasury*

THE MIDLAND LINES FROM LEEDS • 65

Above: **A view** of Bolton Abbey station taken from the south in 1962 shows stored Gresley artic coaching stock in the foreground on the stump of the line to Hambleton Quarry. These coaches were retained for special traffic. In the background can be seen a brick building; this was used by the army during the Second World War, when the yard was used to store ammunition. Bolton Abbey station opened on 16 May 1888, when the line opened to passenger traffic from Skipton. Services were extended through to Ilkley on 1 October 1888, and the station's location – close to the Duke of Devonshire's Bolton Hall estate – ensured that it was frequented on a number by members of the royal family. The last occasion on which the royal train visited the station was in 1947. The MR line from Ilkley to Skipton was authorised by an Act of 16 July 1883. Following closure in 1965, the station's condition deteriorated but, following the site's acquisition by the Embsay & Bolton Abbey Steam Railway in 1995 it has been restored, being officially reopened by Sir William McAlpine on 1 May 1998. *F.W. Smith/Transport Treasury*

Opposite above: **On 11 September** 1962 the photographer has taken advantage of the train's windows to capture Calverley & Rodley station from the east. Although the Leeds & Bradford Railway opened on 1 July 1846, it was not until some date between 16 and 30 July 1846 that the station here opened. Known as Calverley (and for a period as Calverley Bridge), the station was officially renamed Calverley & Rodley on 1 October 1889. The station recorded here was the result of the quadrupling of the line between Leeds and Shipley during the first decade of the twentieth century. Passenger services were officially withdrawn from the station on 22 March 1965, while freight facilities ceased on 7 October 1968. Since the station's closure, the line has been reduced to double track only and the station demolished. The goods shed, however, survives, integrated into a commercial unit. *Henry Priestley/Transport Treasury*

Opposite below: **In July** 1963, a two-car Metro-Cammell-built DMU (later Class 101) is pictured arriving into Skipton station from the east headed by Driver Trailer Composite Lavatory No E56055. During a career that spanned more than forty years – from construction in 1956 through to withdrawal in June 2000 – No E56055 (latterly numbered 54055) was to see service in the north-east, East Anglia and the north-west before ending its main-line days allocated to Longsight. Following withdrawal, No 56055 was preserved and is at the time of writing based on the Cambrian Heritage Railways. *Phil Tatt/Online Transport Archive*

THE MIDLAND LINES FROM LEEDS • 67

Above: As **0-6-0T** 'Jinty' No 47427 shunts the yard in the foreground, Stanier Class 5 4-6-0 No 44901 is seen departing from Skipton with the 4.55pm service from Leeds to Morecambe on 12 June 1964. This view shows to good effect the curvature of the platforms at the station; the track serving the platforms on the route towards Embsay will ascend towards the east and cross over the main line to Keighley slightly to the east of the station. *Alec Swain/Transport Treasury*

Opposite above: **On 4 August** 1964, English Electric Type 3 (later Class 37) No D6772, which was less than two years old when recorded here, heads east at Embsay with a train of tanks from Heysham to Tees Yard. Immediately behind the locomotive is a brake tender, whilst loaded ballast wagons are visible in the sidings on both the Up and Down side awaiting despatch. Apart from a period between January 1970 and December 1972, the locomotive was allocated to Thornaby for its entire operational life and so was returning towards home when pictured. *F.W. Smith/Transport Treasury*

Opposite below: **Slightly to** the east of Embsay station was a private siding that served the quarry operated by the Skipton Rock Co. The final traffic on the Ilkley to Embsay Junction section was the stone traffic that was generated from the quarry. When this ceased in October 1968, this final mile-long section was closed completely and, on 6 July 1969, Embsay Junction signalbox was formally abolished. Thereafter, the line from Skipton to Rylstone was operated under the one train only regulations. *F.W. Smith/Transport Treasury*

THE MIDLAND LINES FROM LEEDS • 69

During 1965, Standard 4-6-0 No 75041 is pictured having brought a short freight train down from the north as it awaits the signal to proceed onto the main line. The original station in Skipton, opened by the Leeds & Bradford Extension Railway on 7 September 1847, was located about quarter of a mile south-east of its current location; the new station was opened on 30 April 1876 to the designs of the MR's architect Charles Trubshaw in connection with the construction of the MR's new route to Carlisle via Settle. The new station had four platforms. The station was further expanded in 1888 when two additional platforms were added for the opening of the line via Embsay to Skipton. These new platforms had originally been provided with awnings although, by the date of this photograph, these had been removed (although the supports are still extant and visible in the view). Trubshaw's station remains and is now listed Grade II. *A W V Mace/Transport Treasury*

The fate of so many of the station on lines covered by this volume; this is the sad sight of Kirkstall station being demolished. Opened originally by the Leeds & Bradford Railway during the second half of July 1846, Kirkstall station was rebuilt by the MR as a result of the quadrupling of the Leeds to Shipley route. The new four-platform station opened on 5 July 1905. Passenger services were withdrawn on 22 March 1965. *Transport Treasury*

Left: **On 29 July** 1966, Standard Class 4MT 4-6-0 No 75011 approaches the Swinden Lime Works at Rylstone with the pick-up freight from Skipton to Grassington. Following the closure of the section north of Rylstone in 1969, traffic along the branch was to become exclusively limestone from the quarry, with the line remaining operational – one of a handful of freight-only routes to survive in the area at the time of writing. *Mike Mitchell/Transport Treasury*

Below: **On 6 August** 1966, the 1.35pm service from Leeds to Morecambe is seen approaching Newlay station. The station, which was known as Newlay & Horsforth from 1 October 1889 until 12 June 1961, had closed on 22 March 1965 and the station buildings visible had been demolished by March the following year. *Mike Mitchell/Transport Treasury*

Above: **The original** Menston station, on the line from Apperley Junction to Burley-in-Wharfedale, opened on 1 March 1873 some eight years after the opening of the route through to Ilkley. The station, known as Menston Junction, was effectively replaced by a new station, illustrated here on 9 March 1968, located 34 chains to the south on 1 November 1875, although until March 1876 both stations were used. The main station building, which, after a period of disuse, underwent restoration to use in 2000 as part of a programme of improvements to the Wharfedale line, is situated on the Down side. To the south of the station, there was a half-mile branch to the west that served High Royds hospital; this was operational between 1883 and 1951 and, although designed primarily to carry freight, carried passengers occasionally free of charge (despite lacking official authorisation to do so). *Geoff D. Smith/Online Transport Archive*

Opposite above: **It is** 9 March 1968 and some three years after the closure of the line towards Otley, the track has been lifted through Arthington station. Although the station itself closed with the line and passengers have not made use of the platforms since March 1965, the station still looks in reasonable condition with much of the signage still extant. Subsequently the station and platforms were demolished. The NER line from Arthington to Otley was authorised by an Act of 11 July 1861 and opened to passenger traffic on 1 February 1865 and to freight traffic on 1 October 1866. *Geoff D. Smith/Online Transport Archive*

Opposite below: **On 9 March** 1968, a somewhat careworn Type 2 (later Class 25) No D5174 stands in the Angle Sidings set within the triangle at Shipley. The sidings were, at one time, used for the storage of coal wagons; the local gas works was not directly connected to the railway network and so coal was delivered to the station, from where it was collected by horse and cart for onward shipment to the works. When this ceased, the sidings were used by permanent way wagons. Built at Darlington and entering service in March 1962, No D5174 (latterly No 25174) was amongst the earliest of the type to be withdrawn (in January 1976). When recorded here, the locomotive was allocated to Holbeck, where it was based between October 1966 and August 1972. *Geoff D. Smith/Online Transport Archive*

THE MIDLAND LINES FROM LEEDS • 73

The terminus at Grassington recorded on 10 March 1968 towards the end of the station's life. The station building survives more than thirty years after the cessation of passenger traffic as does the nameboard, although the latter has clearly seen better days. Freight traffic on the section north of Rylstone – one mile to the south of the terminus – was withdrawn on 11 August 1969 and the final train – an excursion – operated nine days later. Following closure, the station was demolished and a housing estate built on the site. *Geoff D. Smith/Online Transport Archive*

Keighley Junction signalbox viewed in the Down direction on 22 June 1968. The box, which was an MR Type 2A, was listed Grade II in 1986. Redundant and slightly relocated as a result of the main line's electrification in 1994, the condition of the box deteriorated. However, in 2018 Network Rail gained planning consent to dismantle the box and transfer it to Irlam, near Salford, where it was rebuilt and restored; with work completed, it now forms part of a railway display at the station. *Henry Priestley/Transport Treasury*

THE MIDLAND LINES FROM LEEDS • 75

Left: **On 26 July** 1969, as a Bradford Corporation trolleybus passes the closed ex-GNR Shipley station in the background en route towards Saltaire along the A657, a DMU approaches Shipley station with a service towards Bradford Forster Square. The signalbox alongside the train is Leeds Junction; this box was closed on 20 July 1975 following rationalisation of the track layout between Guiseley Junction and Leeds Junction on 11 May 1975. *R.B. Parr/John Meredith Collection/Online Transport Archive*

Below: **Arriving from** the north, Class 25 No 5168 has charge of a short train of scrap metal as it approaches Bingley station on 1 September 1972. The section of the line from Shipley to Keighley, authorised as the Leeds & Bradford (Shipley-Colne Extension) Railway, opened on 16 March 1847. When the line first opened, passenger services operated from Bradford to Skipton; passengers for Leeds changed at Shipley. Visible in the background is the goods shed; freight traffic was handled until 28 June 1965. The goods shed remains, however, and is now in industrial use. The contemporary view from here is, however, very different. Although the railway is still open and now electrified, the land between the railway and the Leeds-Liverpool Canal has been utilised for the new A650 Airevalley Road (the Bingley bypass); the work required a new cut for the canal through Bingley at this point with the new road opening in 2003. *John Meredith/Online Transport Archive*

On the same day, a two-car set, comprising Class 104 Nos E50596 (leading) and E56187, stands in the Up platform at Bingley with a service from Skipton to Leeds. Although Bingley opened, as the only intermediate station on the line from Shipley to Keighley initially, on 16 March 1847, the station as seen here was the result of a relocation, with the new station – designed by Charles Trubshaw – being opened on 24 July 1892. The original station was located about 400 yards to the north adjacent to the Leeds-Liverpool Canal. The Trubshaw-designed station is now Grade II listed. *John Meredith/Online Transport Archive*

On 7 June 1973, a two-car DMU forms the 8.45am service from Ilkley to Bradford, which is seen departing from Burley-in-Wharfedale station. The train is passing the then redundant goods shed – freight facilities were withdrawn from the station on 27 April 1964 – and the building was to be demolished shortly after the date that this photograph was taken. *F.W. Smith/Transport Treasury*

Despite its destination on 30 March 1974, this two-car DMU standing in platform 2 at Shipley station is actually heading towards Bradford. The station at Shipley was rebuilt between 1883 and 1892 to the design of the MR's architect, Charles Trubshaw. Modernisation allied to the sharp curvature of the tracks through platforms 1 and 2, which precluded modern trains being able to pass in the station, resulted in the closure of platform 1. Subsequently, with the opening of the platforms on the direct Leeds to Skipton side of the triangle, the platforms were renumbered and platform 2 was renumbered five. Although the station buildings visible within the triangle are still extant, those on the former platform 1 have been demolished. *Chris Gammell/Online Transport Archive*

On 16 March 1975, Class 40 No 40087 is seen arriving at Keighley wrong line with an Up freight heading towards Leeds. The Leeds & Bradford (Shipley-Colne Extension) Railway line from Keighley to Skipton opened on 8 September 1847; initially only a single track was operational but the second line was in operation by the end of the year. The train is passing Keighley Junction signalbox and the line in the foreground is the connection for the branch – by this date preserved – to Oxenhope. *John Meredith/Online Transport Archive*

Above: **The approach** to Ilkley station from the east, where the through lines into platforms 3 and 4 diverged from the terminal platforms 1 and 2, was controlled by Ilkley Junction signalbox. The box, with its cantilevered design (favoured by the MR when installing boxes in confined spaces), opened in 1913 and was initially equipped with a forty-four-lever Tappet frame. Recorded here in the 1980s, towards the end of its life, the box was latterly equipped with a panel. The box was closed in 1994 and subsequently demolished. *J.G.S. Smith/Transport Treasury*

Opposite above: **On 9 April** 1984, Saltaire station was reopened, almost twenty years after passenger services were withdrawn. This is the scene as the crowds witness the first arrival – the reopening special – formed of Class 141 No 141006 in the then West Yorkshire PTE livery of Verona green and buttermilk. Although the new platforms were constructed from wood, in recognition of the architecture of Saltaire itself – now a Unesco World Heritage site – the platform shelters were constructed in stone. No 141006 – latterly No 141107 – was one of the batch sold to Iran during 2001 and 2002. *Gavin Morrison*

Opposite below: **The ex-MR** Shipley Bradford Junction signalbox viewed from the station on 2 May 1987. The siding on the right served the Crossley Evans scrapyard. The box, an MR Type 3B design, accommodated a thirty-six-lever frame and was opened on 27 September 1903 to replace an earlier structure. A new signalling panel was installed on 21 October 1984; this was modified on 11 June 1990 in connection with the relocation of Forster Square station. The box closed on 25 June 1994, when control of the section passed to the power box at Leeds, and the box itself was demolished four days later. *Author*

THE MIDLAND LINES FROM LEEDS • 79

Above: **On 2 May** 1987, Class 144 Pacer No 144020 stands in Shipley with a service bound for Bradford Forster Square. When recorded here, No 144020 was virtually brand-new; a total of twenty-three of the class were constructed for services within the West Yorkshire MCC area during 1986 and 1987 and were painted in the council's crimson and cream livery. Originally all twenty-three were two-car sets as illustrated here but subsequently ten units – Nos 144014-023 – were converted to three-car. When delivered, the class was notoriously unreliable in service, often requiring replacement by other multiple units or locomotive-hauled sets; these problems – the mechanical transmission and wheel slippage – were largely resolved as the result of subsequent modification and the class continued to give service through the era of privatisation – passing through the hands of several franchisees – before withdrawal during 2020 as a result of non-compliance with modern accessibility standards. A significant number – including No 144020 (based on the Wensleydale Railway at the time of writing) – have survived into preservation. *Author*

Opposite above: **In 1995,** shortly before the introduction of electric services, two Class 144 Pacers – Nos 144009 (two-car) on the left 144014 (three-car) on the right – stand in Ilkley station with departures for Leeds and Bradford respectively. All the Class 144s were eventually withdrawn by mid-2020; this was slightly later than planned as a result of the late delivery of the replacement units (Classes 195 and 331). After withdrawal, eighteen of the sets were stored temporarily on the Keighley & Worth Valley Railway with the remaining five at Heaton depot pending disposal. A number have been preserved; three sets – including No 144014 – have been acquired by Vintage Trains for use on the main line whilst No 144009 is, at the time of writing, stored on the East Lancs Railway pending transfer to the Greater Manchester Fire & Rescue Services for use in training. Whilst based on the East Lancs, it was painted in a spurious Great Midlands Trains livery for use in filming. *Geoffrey Tribe/Online Transport Archive*

Opposite below: **The original** Saltaire station closed on 22 March 1965 and was demolished five years later. On 9 April 1984 a new station – promoted by the West Yorkshire Passenger Transport Executive and British Rail – opened on the original site. The new station was provided with wooden platforms although – unusually and in keeping with the World Heritage Site that the station serves – the platform shelters were constructed in stone. In 1995, Class 156 DMU No 156472 approaches the station with a service heading towards Morecambe. *Geoffrey Tribe/Online Transport Archive*

THE MIDLAND LINES FROM LEEDS • 81

Above: **Although work** on the electrification of the lines from Leeds to Bradford Forster Square, Ilkley and Skipton was undertaken during the early 1990s, no new rolling stock was ordered for the electric services. As a consequence, Regional Railways took over the Class 308/1 EMUs that had been built at York for use on the Great Eastern electrification. These were overhauled at Doncaster Works and reduced from four to three carriages at the same time. One of the three-car sets – No 308161 – is seen here at Keighley on 1 July 1996 at Keighley with a service towards Skipton. All of the Class 308/1s were withdrawn by the end of 2001, being replaced by Class 333s. *Les Folkard/Online Transport Archive*

Opposite above: **In June** 2000, one of the Class 308 EMUs transferred to the area for operation over the electrified lines to Skipton and Ilkley is pictured in the new platform 1 at Shipley station with a Leeds-bound service. When Shipley station was rebuilt in the late nineteenth century, it lacked platforms on the main line. This omission was rectified in two stages: firstly, a single platform on the Down line was opened in May 1979, with Up services forced to cross over to the Down line to make use of it, with the platform on the Up line following in 1992. *Vintage Carriages Trust/Online Transport Archive*

Opposite below: **Baildon station** originally opened with the completion of the MR line from Shipley Guiseley Junction to Esholt Junction; it was one of two stations on the route – the second, Esholt, closed on 28 October 1940. Baildon station closed on 5 January 1953; however, it briefly reopened between 28 January and 29 April 1957 as a result of the Suez crisis when petrol supplies were restricted and rationing was introduced. Although the Bradford to Ilkley service was threatened by Beeching, the line survived and – following a long campaign and as a result of financial support by Baildon Urban District Council and Bradford Corporation – passenger services were restored on 5 January 1973. Originally the reopened station was provided with two platforms, but the line was singled in 1983 with only the former Ilkley-bound platform being retained. The redundant track was lifted the following year. On 30 June 2010, Class 333 No 333002 is pictured arriving at the station with a service towards Ilkley. *Gavin Morrison*

THE MIDLAND LINES FROM LEEDS • 83

Located between Bingley and Keighley, the new station at Crossflatts was the first of the new or reopened stations on the Airedale route. Opened on 17 May 1982, the station, is seen here on 30 June 2010 as two Class 333 EMUs pass. Heading eastbound with the 15.00 service from Skipton to Bradford Forster Square is No 333014. *Gavin Morrison*

The original station at Apperley Bridge was one of those closed on 22 March 1965 following the Beeching report. It had opened on 30 April 1846, when it was provided with two platforms, but with the quadrupling of the line at the start of the twentieth century, a further two platforms were constructed. Following closure, the station was demolished. In 1999, West Yorkshire PTE announced that the station was one of five that the authority would like to see reopened over the next five years; in the event, however, it and Kirkstall Forge were not to be progressed for a number of years. Construction was sanctioned in the National Infrastructure Plan of 29 November 2011 and the new two-platform station, on a slightly different site to the original, at Apperley Bridge finally opened on 13 December 2015. On 1 June 2021, Class 333 No 333014 is pictured calling at the new station with a westbound service. *Gavin Morrison*

HALIFAX AND THE CALDER VALLEY

The Stainland branch, with its intermediate stations of Rochdale Road Halt (opened on 1 March 1907) and West Vale (situated across the road from where the author's father grew up), was authorised by Acts of 5 July 1865 and 16 July 1874. It opened for passenger traffic on 1 January 1875 and to freight on 29 September 1875. However, the branch's finances were undermined by the arrival of Halifax Corporation's electric tramcars and during the 1920s by the deterioration of the country's economy with the result that passenger services were withdrawn on 23 September 1929. Freight traffic, serving the yards at West Vale and Stainland for the many mills in the valley, continued until 14 September 1959. On 5 May 1951 the branch was visited by a joint SLS/MLS special – the 'Pennine Rail Tour' – and the train is pictured at Stainland behind ex-LYR 2-4-2T No 50865. *Tony Wickens/Online Transport Archive*

Above: **On 30 July** 1951, 2-6-4T No 42189 draws into Greetland station with a service to Bradford Exchange via Halifax. Designed by Bradford-born (and educated) Charles Fairburn, No 42189 must have been amongst the last LMS-built locomotives to have been completed prior to the nationalisation of the railways on 1 January 1948 as it emerged from Derby Works on 24 December 1947. Allocated to Low Moor when new, it was based at Manningham between September 1960 and May 1967 when it was transferred to Wakefield. A further transfer two months later saw it move to Normanton, from where it was withdrawn in September 1967. Greetland station, with the junction for the Stainland branch visible in the background, lost its passenger services on 10 September 1962. *Tony Wickens/Online Transport Archive*

Opposite above: **On 8 August** 1951 'Austerity' 2-8-0 No 90697 is recorded departing from the yard at Pellon heading towards Holmfield. The steeply-graded and heavily engineered 2½-mile long line was one the most challenging for locomotive crews in the area. Originally constructed with double track, passenger services were only offered by the GNR although both the GNR and LYR provided freight trains. The line – as illustrated here – was singled and lost most of its signalling; however, its proximity to a significant number of mills ensured its survival for more than forty years as a freight-only branch following the withdrawal of passenger services as a wartime measure during the First World War. *Bryan Jennings*

Opposite below: **With the** closed station of Cooper Bridge in the background, ex-LYR Class 5P 4-6-0 No 50455 heads west. A total of seventy-five of the type – the largest passenger locomotives designed for use by the LYR and nicknamed the 'Lanky Dreadnoughts' – were constructed between 1908 and 1925 to the design of George Hughes, although the earlier locomotives, built in 1908 and 1909, were not a success and were subsequently rebuilt. The last twenty of the type constructed, including No 50455, were originally planned as tank locomotives and had a slightly different wheelbase. Of the seventy-five, only seven passed to BR in 1948. Of these, only one – No 50455 – was to receive its BR number; when recorded here, it was approaching the end of its operational life as it was withdrawn in October 1951. When the Manchester & Leeds main line east of Hebden Bridge first opened in 1840, the station at Cooper Bridge was designed to serve Huddersfield (which at that stage had no station and was some four miles away). The station closed on 20 February 1950 and was subsequently demolished although there remain traces of the access to the platforms in the road underbridge to the west of the station over which the train is passing. *Neville Stead Collection/Transport Treasury*

HALIFAX AND THE CALDER VALLEY • 87

Above: On St George's Day 1952 – 23 April – Class N1 69478 departs from North Bridge station with a service towards Queensbury. Jointly promoted by the Great Northern and Lancashire & Yorkshire railways, the Halifax & Ovenden Joint was opened from Halifax Old to North Bridge for freight traffic on 17 August 1874 and thence to Holmfield a fortnight later. Passenger services were not introduced until 1 December 1879. The latter ceased on 23 May 1955 and the line northwards from North Bridge closed completely on 27 June 1960. The remaining section – to serve the goods yard at North Bridge – survived until 1 April 1974. Subsequently the viaduct that linked the two stations in Halifax was demolished. *Bryan Jennings*

Opposite above: On the same day, Class J50 No 68944 heads into North Bridge with a – very – short freight. The line to the right of the locomotive formed a connection into the large Dean Clough carpet mills owned by the Crossley family whilst the tunnel and huge retaining walls emphasise the determination of the Victorian engineers to take the railway through the often difficult terrain encountered in the region. Like many Victorian entrepreneurs, Francis Crossley – the then owner of Dean Clough Mills – acquired a country estate; he purchased Somerleyton Hall, near Lowestoft, in 1863 from the entrepreneur and railway contractor Samuel Morton Peto, who was facing financial ruin at the time and went bankrupt three years later. *Bryan Jennings*

Opposite below: Another freight-only branch visited by the joint MLS/SLS 'West Riding' rail tour on 6 September 1953 was the section from Holmfield to Halifax St Paul's; the train is pictured at the terminus. The branch from Holmfield to St Paul's opened for freight in two stages: from Holmfield to Pellon on 1 August 1890 and thence to St Paul's on 5 September 1890. Passenger services operated over the branch also commenced on that day, but were never a great financial success. Competition from the corporation's tramways allied to wartime exigencies resulted in passenger services being withdrawn on 1 January 1917. The Halifax High Level line closed completely on 27 June 1960. The station was subsequently demolished and the site redeveloped. *Tony Wickens/Online Transport Archive*

HALIFAX AND THE CALDER VALLEY • 89

Above: **With the** coal-fired Thornhill Power Station forming the backdrop, Class 5 4-6-0 No 45037 departs from Ravensthorpe & Thornhill station. Although the main line from Dewsbury Junction (Thornhill) to Leeds opened on 18 September 1848, it was not until 1 September 1891 that the station at Ravensthorpe & Thornhill opened. The station lost its '& Thornhill' suffix in 1959. The buildings at the station, which remains open, were demolished following fire damage and replaced by simple platform shelters. Thornhill Power Station, which was rail served (as evinced by the coal wagons visible), was constructed for the Yorkshire Electric Power Co and opened in 1902. As pictured here, the facility had undergone a major upgrade between 1950 and 1954. The power station closed in 1982 and was subsequently demolished; in 1998 a smaller, gas-fired, power station was opened on the same site. *Neville Stead Collection/ Transport Treasury*

Opposite above: **On 12 July** 1959, 'Jubilee' class No 45717 *Dauntless* passes through Brighouse station with the 10.30am service from Liverpool Exchange to Newcastle. When the station opened, courtesy of the Manchester & Leeds Railway, on 5 October 1840 it was known as 'Brighouse for Bradford' as Bradford was not as yet served by the railway. The station illustrated here, however, was opened on 1 May 1893 when the station was relocated about 350 yards to the west. Although not scheduled for closure in the Beeching Report, declining traffic resulted in the station's demise on 5 January 1970, although the line remained operational primarily for freight traffic thereafter. *Gavin Morrison*

Opposite below: **Seen departing** from Halifax with a westbound service on 15 July 1959 is Fowler-designed 2-6-4T No 42409. The original station at Halifax was opened on 1 July 1844 when the Manchester & Leeds Railway opened its branch from Sowerby Bridge; this station was sited at Shaw Syke, about 200 yards south-west of the station illustrated here. The first station was closed on 7 August 1850 with the opening of the line to Low Moor; it was then converted and extended for use as a goods yard. The main station building at the new station, which is now Grade II listed, was designed by Thomas Butterworth but the platforms as illustrated here were the result of a remodelling undertaken during 1885 and 1886 to accommodate GNR services from Bradford and Keighley via Queensbury. Between 1899 and 30 September 1951, the station was known as 'Halifax Old' in order to differentiate it from Halifax North Bridge on the Queensbury line; it was renamed Halifax Town on 30 September 1951, a name it carried until it reverted simply to Halifax, following the closure in 1955 of North Bridge, on 12 June 1961. *Gavin Morrison*

HALIFAX AND THE CALDER VALLEY • 91

Above: **On 1 June** 1960, less than a month before the final closure of the Halifax High Level branch, Austerity 2-8-0 No 90122 stands in front of the substantial goods shed at Pellon. When the High Level line first opened for freight traffic from Holmfield, on 1 August 1890, the original terminus was at Pellon. It was not until 5 September 1890 that the route was completed through to St Paul's and passenger services introduced. The passenger service over the branch also commenced on 5 September 1890, but this was destined to be relatively short-lived as a journey from St Paul's to the town's main station – a distance of less than two miles – took around forty minutes and involved a change at Holmfield generally. The arrival of the competing tramcars, operated by Halifax Corporation, saw passenger services over the High Level branch withdrawn on 1 January 1917. *Gavin Morrison*

Opposite above: **Following the** withdrawal of passenger services over the Queensbury Triangle routes to Bradford Exchange and Keighley in May 1955, the section between Holmfield and Queensbury closed completely. Freight traffic, however, continued to be handled on the line from Halifax to Holmfield, where there was an extensive freight yard to see the numerous adjacent textile mills, and down the High Level route to Halifax St Paul's. The freight traffic was to survive for a further five years, with the yards at St Paul's, Pellon and Holmfield all being officially closed on 27 June 1960. Two days after this, Class 5 No 45339 is seen at Holmfield collecting the last wagons from the yard. After closure, the station was demolished and the site redeveloped for commercial and industrial use. *Gavin Morrison*

Opposite below: On 30 September 1961, Austerity 2-8-0 No 90376 approaches Greetland from the east with a freight. In the foreground is the junction for the branch to Stainland. By the date of the photograph, the branch has finally been closed although the track was used for the storage of withdrawn wagons for a period thereafter until being lifted. *Neville Stead Collection/Transport Treasury*

HALIFAX AND THE CALDER VALLEY • 93

Above: **Although the** main line through Mirfield opened, courtesy of the Manchester & Leeds Railway, on 5 October 1840, it was not until April 1845 that the first station to serve the town opened. The station as it exists today, however, is largely the result of work undertaken during the 1860s, when the station was relocated 202 yards to the east. Contracts for the new station and overall roof were let on 25 May 1864 and 26 April 1865 respectively, with the new station, with its large island platform, opening on 5 March 1866. One hundred years later, in 1966, 'Jubilee' class No 45562 *Alberta* is pictured awaiting departure with a westbound service. Although Mirfield station remains open, its importance declined with the closure of the line through the Spen Valley to Bradford on 14 June 1965 and the withdrawal of passenger services over the erstwhile main line beyond Normanton to York on 5 January 1970. The decline of the station resulted in the existing platform buildings being demolished; the overall roof was dismantled during 1977. *Neville Stead Collection/Transport Treasury*

Opposite above: **The station** at Elland originally opened on 5 October 1840 but was relocated 607ft to the east on 1 August 1865. Freight facilities were withdrawn on 28 June 1962 with closure to passenger traffic following less than four months later. On 16 July 1966, Class 5 No 44732 is seen passing the site of the now demolished station with an Up service towards Manchester. Visible in the distance is the signalbox; this was a sixty-lever box that was constructed as a replacement by BR in 1958. The box was finally closed in 2008 and subsequently demolished. There were plans to reopen Elland contemporaneously with Brighouse in 200l; however, lack of funds precluded this. In 2014, a study indicated that there was a strong business case for its reopening. In June 2017, funds for this were allocated by the West Yorkshire Combined Authority and a planning application for the work was lodged in 2021. *Gavin Morrison*

Opposite below: **Seen approaching** Thornhill Junction from the west on 30 July 1966 with the 8.48am service from Halifax to King's Cross is Fairburn-designed 2-6-4T No 42077. The origins of the main line through Thornhill lay with the Manchester & Leeds Railway, which opened its route from Hebden Bridge through to Goose Hill Junction on 5 October 1840. *Mike Mitchell/Transport Treasury*

HALIFAX AND THE CALDER VALLEY • 95

Above: **On 27 July** 1968, English Electric Type 3 (later Class 37) No D6925 is pictured light engine at Normanton. The station at Normanton was opened on 30 June 1840 by the North Midland Railway; its importance grew as it was later served by the trains of the Manchester & Leeds and the York & North Midland railways. In 1842, the Normanton Hotel and Refreshment Rooms opened; this was linked to the station by a footbridge and long-distance passenger services paused at the station to permit travellers to take refreshment. Normanton's pre-eminence was not, however, to last as Anglo-Scottish services over the East Coast main line were diverted over the line from Doncaster to York and by the completion of the LNWR route across the Pennines and the Leeds & Thirsk line, which offered an alternative route from the Manchester area to north-east England. The reduction of the railway network during the final decades of the twentieth century saw the importance of Normanton further reduced and the contemporary station, with the railway now no more than a two-track section, serves only an island platform. *Neil Caplan/Online Transport Archive*

Opposite above: **On 25 May** 2010 – the first day of the company's operations – Grand Central Class 180 No 180107 is seen at Brighouse station with the 12.08 service from Bradford Interchange to King's Cross. Some thirty years after the original station closed, Brighouse returned to the railway network when the new station was opened on 28 May 2000. Initially, the station was served by hourly trains between Leeds and Huddersfield that operated via Halifax and Bradford Interchange, but these were supplemented in 2007 by a service from Leeds to Southport via Brighouse, Hebden Bridge and Manchester Victoria. Grand Central is one of the Open Access TOCs that arose post-railway privatisation. The company's initial services were between Sunderland and King's Cross but in March 2008, it sought permission to operate services from Bradford to London; these were scheduled to commence operation in late 2009 but problems in the delivery of the stock delayed this until May 2010. In November 2011, Grand Central became a subsidiary of Arriva (itself now a subsidiary of DB). *Gavin Morrison*

Opposite below: **On 6 June** 2010, Grand Central Class 180 No 180112 is pictured departing from Halifax with the 14.39 Saturdays Only service from King's Cross to Bradford Interchange. The fourteen-strong Class 180 was constructed by Alstom at Washwood Heath, Birmingham, for use by First Great Western from 2001; however, operational problems resulted in the TOC returning them to the leasing company in 2008 (although five were to return for a period). Subsequently, Grand Central leased two for its initial services; a further three were acquired with the start of the Bradford service and the operator also took over the five ex-First Great Western units in 2017. The remaining four were leased by Hull Trains, another Open Access TOC, and then by East Midlands Railway. Visible is Halifax signalbox; this had originally been Halifax East and was opened by the LYR in 1884. The Grade II listed box remained in operational use until 20 October 2018 when, along with closure of the boxes at Hebden Bridge, Milner Royd Junction and Mill Lane, the York Rail Operating Centre assumed control the area. *Gavin Morrison*

HALIFAX AND THE CALDER VALLEY • 97

Above: **Pictured heading** west with the Wilton to Collyhurst binliner train on 8 August 2018 is Class 66 No 66152 in DB Schenker livery. The train is passing through the site of the long-closed station at Thornhill. When the station here opened on 5 October 1840, it was known as Dewsbury before being renamed (over the years a number of variations were recorded including Thornhill for Dewsbury). The station closed on 1 January 1962. *Gavin Morrison*

Opposite above: **On 24 July** 1976, a Class 124 TransPennine DMU approaches Mirfield from the east with an Up service via Huddersfield towards Manchester and Liverpool. The distinctive DMUs, more stylish than some of the older InterCity/Cross Country sets as a result of the involvement of E.G. Wilkes in their design following criticism of some of the early units produced, were built at Swindon during 1960 and 1961. Initially, eight six-car sets plus three spare vehicles were constructed, with two of the non-driving vehicles in each set being equipped with motors in order to deal with the gradients on the route. Services were launched on 29 December 1960 with the Hull-Liverpool Lime street service being formally introduced on 2 January 1961. The six-car sets included a buffet car; however, declining traffic resulted in the first four of the eight buffet cars built being withdrawn in 1971 and five-car sets became increasingly the norm, as illustrated here. Although one of the four withdrawn buffet cars was reinstated when the buffet service was revamped, all five of the surviving buffet cars were withdrawn in 1975. From 1977, the transfer of ex-Western Region Class 123s resulted in the operation of hybrid units. The first casualty, No E51959, was scrapped as a result of an accident in October 1977, but all were withdrawn by the end of 1984. Although the North Yorkshire Moors Railway was interested in the preservation of a three-car set, this came to nothing as the result of issues over asbestos removal and, as a result, none of the type survives in preservation. *Gavin Morrison*

Opposite below: **In connection** with the upgrade to the TransPennine route and the partial electrification of the network over which it operates, TransPennine Express was one of three TOCs that placed orders with Japanese company Hitachi for the supply of bi-modal – diesel and 25kV electric – units based on the design of the Japanese A-train/AT300 design. TransPennine Express ordered nineteen five-car sets; these were originally to be designated Class 803 but, in the event, became Class 802/2 with the brand name Nova 1. Production of the batch commenced in December 2017 and the first units commenced testing in July the following year. The first Class 802/2-operated service ran on 28 September 2019. Here No 802214 is seen passing through Ravensthorpe station in diesel mode with the 10.46 service from Newcastle to Liverpool on 24 June 2020. *Gavin Morrison*

HALIFAX AND THE CALDER VALLEY • 99

THE HEAVY WOOLLEN DISTRICT

The complex network of lines along the Calder Valley corridor as shown in the Railway Clearing House map of 1911. The three stations that served Dewsbury – the through stations on the GNR and LNWR routes and the LYR terminus – are all shown, as is the MR's terminus which was only ever used for freight traffic.

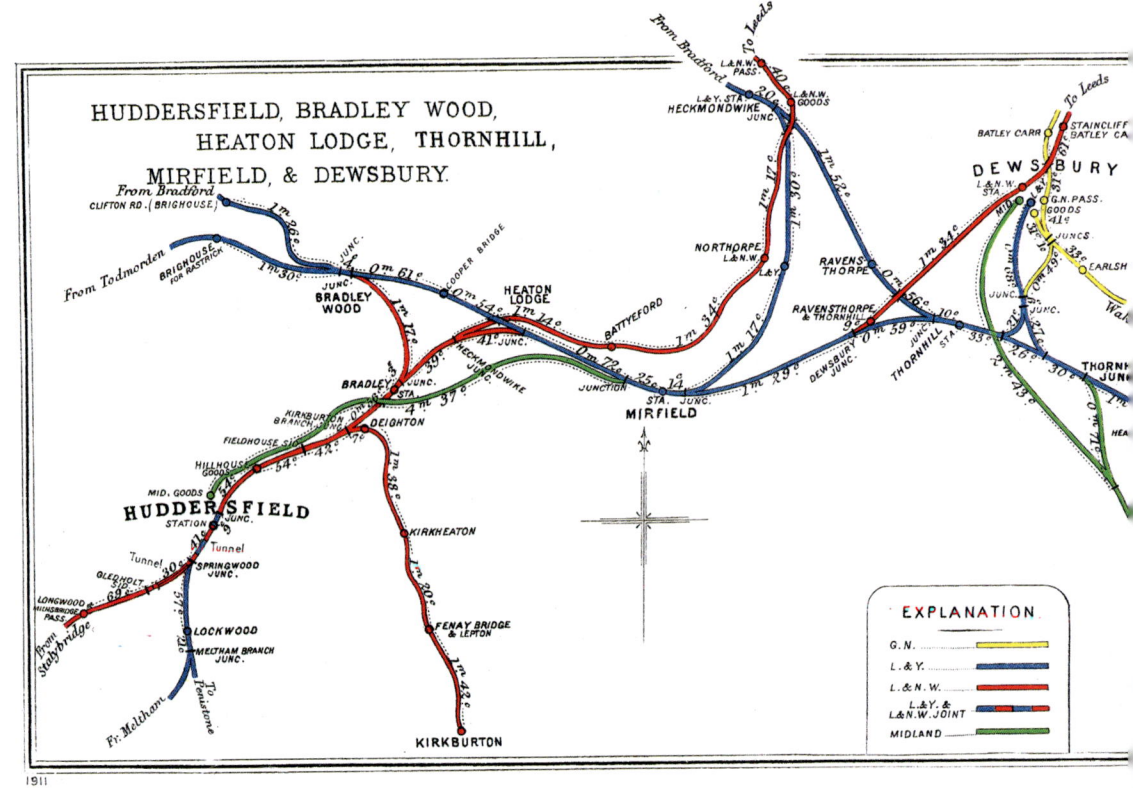

On 28 July 1977, a two-car Class 101 DMU approaches Batley station from the north with a service towards Huddersfield. The line from Thornhill Junction, on the Manchester & Leeds line from Sowerby Bridge to Normanton, was authorised as the Leeds, Dewsbury & Manchester Railway by Acts of 30 June 1845 and 27 July 1846. The original LNWR station here opened, with the line itself, on 18 September 1848. Although not evident in this view, the stone-built single-storey main station building is still extant at Batley, serving the northbound platform. Batley's importance grew with the opening of the LNWR's branch to Birstall (part of the railway's unsuccessful attempt to reach Bradford with its own direct route) along with the completion of the GNR's Dewsbury loop and the opening of line to Tingley. The station was rebuilt in 1890 with an additional three platforms; these were demolished following the withdrawal of the Bradford to Wakefield via Batley service on 7 September 1964. Since this view was recorded, the area to the east of the station has been redeveloped. *Gavin Morrison*

The LNER timetable for the period from 16 June to 5 October 1947 covering the lines from Bradford Exchange to Wakefield via either Ardsley or Batley.

Above: **Dewsbury Central** was the GNR's station in the town and opened on 12 April 1880, with the extension to Batley, to replace an earlier station that dated to 9 September 1874. The running-in board includes the suffix 'Central'; this was a relatively recent renaming when Class N1 0-6-2T No 69452 was recorded on a Wakefield train on 30 July 1951. Passenger services on the line from Wakefield to Bradford Exchange via Dewsbury were withdrawn on 7 September 1964. *Tony Wickens/Online Transport Archive*

Opposite above: **The end** of a dream – the buffer stops at Thornhill that marked the end of the proposed cut-off line from Royston Junction via Middlestown Junction to Bradford on 30 July 1951. The MR line north from Royston Junction was authorised of 25 July 1898 and 13 July 1899; it opened from Royston Junction to Crigglestone for freight traffic on 1 July 1905 and thence to Thornhill Junction, on the LYR main line, on 10 November 1906. Passenger services were introduced over the route on 1 July 1909. Where once it was planned that services to and from Scotland via the Settle & Carlisle line would have thundered, the track was in use solely for the storage of carriages. The line between Thornhill and Royston Junction was largely freight only from 1946 – except for a brief revival in 1960 – and was to close completely during 1968. Amongst the traces that survive from the route is the twenty-one-arch Grade II listed Crigglestone Viaduct, which stretches for a distance of 1,270 feet over the line from Wakefield to Barnsley. *Tony Wickens/Online Transport Archive*

Opposite below: **Originally the** Midland Railway planned on opening a passenger station in Dewsbury on the through route that it proposed to construct between Royston Junction and Bradford; in the event, the MR opened a connection from Middlestown Junction to Thornhill and operated through services from Sheffield to Halifax via the LYR line west from Thornhill. The section from Middlestown Junction through to Dewsbury, which was authorised by Acts of 30 July 1900 and 30 June 1903, was opened – to freight traffic only – on 1 March 1906. The MR goods yard – Savile Town – was accessed off the proposed cut-off route to Bradford by a short branch that descended at a 1 in 40 gradient through a 188-yard tunnel. Savile Town closed on 18 December 1950 although when seen here on 30 July 1951 the track and shed were still intact. The largely railway-owned commercial vehicles – of a variety of interesting marques (including a Vulcan) – seem out of use and presumably awaiting disposal. *Tony Wickens/Online Transport Archive*

THE HEAVY WOOLLEN DISTRICT • 103

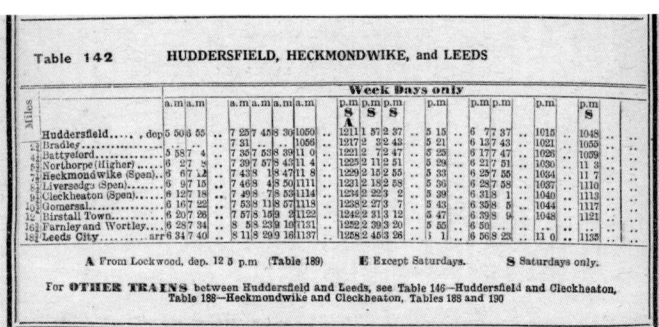

Above: **The passenger** timetable for the 'Leeds New' line operated by the LMS between 16 June and 5 October 1947.

Below: **By the** end of the nineteenth century, the existing line from Huddersfield to Leeds was becoming inadequate for the traffic carried. The existing route was, however, unsuitable for quadrupling and so an alternative route – from Spen Valley Junction to Farnley Junction via Cleckheaton – was promoted. Authorised by two Acts of 27 June 1892 and 6 July 1895, the route, known as the 'Leeds New Line', opened for goods traffic from Spen Valley Junction to Northorpe on 18 September 1899 and thence to Farnley Junction on 9 July 1900. Passenger traffic commenced on 1 October 1900. The first station northwards on the route from Spen Valley Junction was Battyeford and here 2-6-4T No 42310 is pictured at the station with a southbound service from Leeds on 12 September 1963, shortly before the withdrawal of passenger services. Local passenger traffic ceased over the route on 5 October 1953 although the line was used as a diversionary route until 2 August 1965. Freight traffic continued until 11 January 1966 when the entire route – with the exception of the short section from Liversedge to a new connection with the ex-LYR route – was closed completely. *Neville Stead Collection/Transport Treasury*

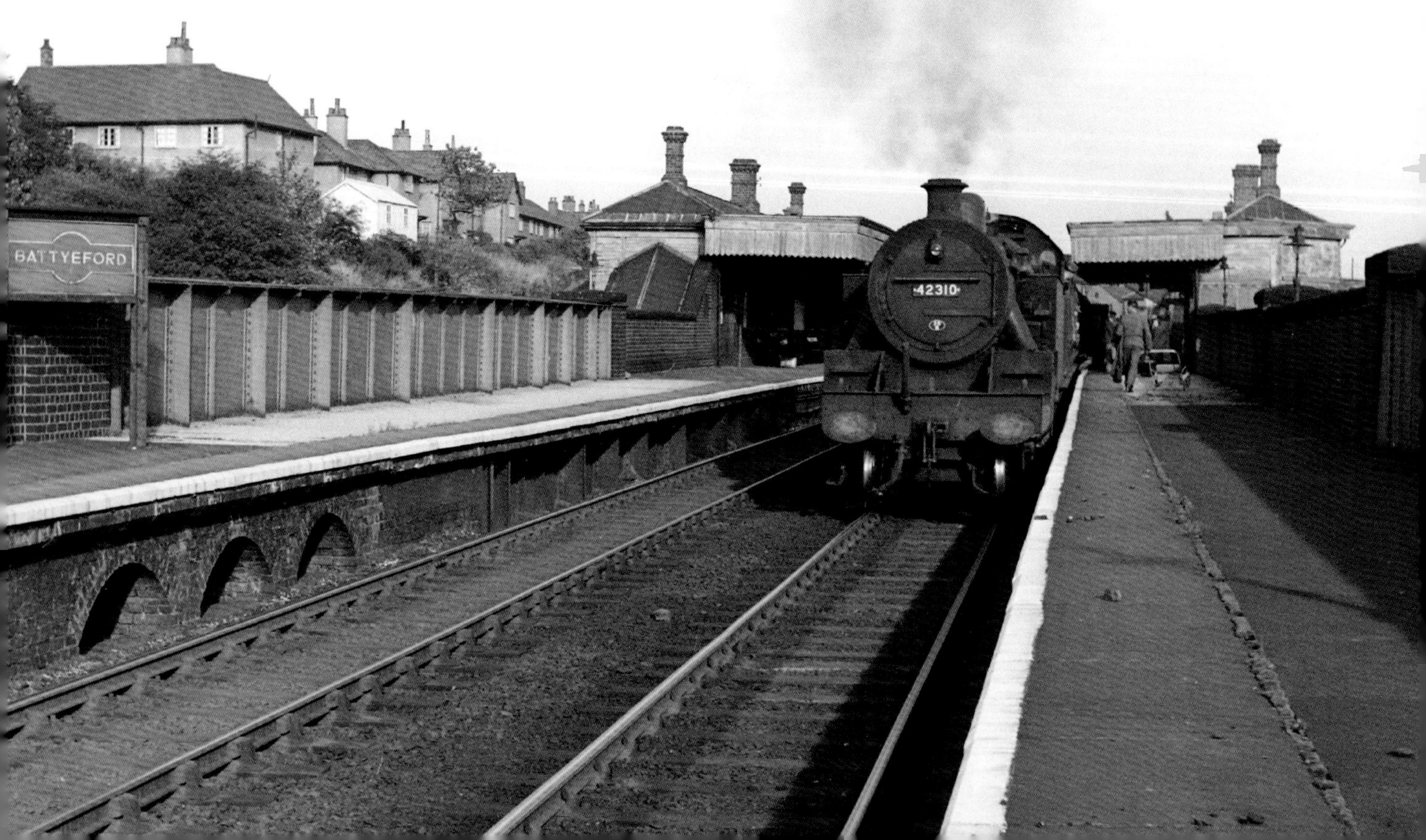

Left: **The opening** of the GNR's Dewsbury loop on 12 April 1880 meant that the existing branch between Ossett and Batley declined in importance and, on 1 July 1909, passenger services between Runtlings Lane Junction and Batley ceased; there was only one intermediate station – Chickenley Heath. The line remained open, however, for freight as it served Shaw Cross Colliery and, on 6 September 1953, the joint SLS/MLS 'West Riding' rail tour headed by Class N1 0-6-2T No 69430 traversed the line; the train is pictured here at the site of the closed Chickenley Heath station. The route was closed completely south of the colliery on 26 March 1956, but the northern section survived until 1 May 1972 for coal traffic. *Tony Wickens/Online Transport Archive*

Below: **In 1960,** Class B1 No 61158 has just departed from Heckmondwike Central with a southbound service. The station at Heckmondwike was relocated slightly to the west on 9 August 1888, when the level crossing adjacent to the original station was replaced by a road underbridge. The first station – opened on 18 July 1848 – was situated alongside the freight yard visible in the foreground. The station acquired the suffix 'Central' on 2 June 1924, retaining it until 12 June 1961. The link from Thornhill to Heckmondwike opened to freight traffic on 10 May 1869 and to passenger services on the following 1 June. The latter were withdrawn on 14 June 1965 with the line to Mirfield closing completely at the same time. Freight facilities were withdrawn from Heckmondwike on 5 May 1969. *Neville Stead Collection*

Above: **'Jubilee' class 4-6-0** No 45649 *Hawkins* is seen on 6 July 1961 passing Gildersome East signalbox with a westbound service as it approaches Gildersome Tunnel. By this date, the ex-LNWR station at Gildersome was long closed – it lasted barely twenty years between opening on 1 October 1900 and closing finally on 11 July 1921 (having been closed for almost two years during the war) – although evidence of the platforms can still be seen. Today the site of the station is sandwiched between the A62 and the M621, with the cutting between the tunnel and the Rooms Lane overbridge now infilled; however, the stone bridge parapets are still extant to illustrate where the line once went. *Gavin Morrison*

Opposite above: **Pictured climbing** away from Batley on 27 July 1963 are two Class 5s – Nos 45219 and 44694 – at the head of the 1.35pm Saturdays Only service from Cleethorpes to Bradford Exchange. The train will ascend to climb over the ex-LNWR main line towards Leeds en route towards Laisterdyke and its destination. The line from Adwalton Junction was extended from Upper Batley to Batley itself on 1 November 1864. Passenger services over the three-mile line from Adwalton ceased on 7 September 1964 with freight traffic being withdrawn on 15 February 1965. The trackbed of the former GNR route is still extant at this point, as are the abutments of the overbridge (adjacent to the Howley Street level crossing) that carried the line over the line to Leeds, but the route towards Adwalton westwards is less well preserved. *Mike Mitchell/Transport Treasury*

Opposite below: **Having just** departed from Batley, 'Jubilee' No 45643 *Rodney* heads the 8.55am Saturdays Only service from Bournemouth to Leeds towards Morley Tunnel. The 4-6-0 had been reallocated to Farnley Junction shed in early November 1963; a final transfer saw the locomotive move the short distance to Holbeck in early October 1965 from where it was withdrawn during the following January. *Mike Mitchell/Transport Treasury*

THE HEAVY WOOLLEN DISTRICT • 107

Above: **The ex-LYR** line from Mirfield to Low Moor via Cleckheaton lost its passenger services from 14 June 1965; two days before that date (the actual last day of passenger operation), Class B1 No 61016 is seen departing southbound from the island platform station. Freight facilities were withdrawn from Cleckheaton on 5 May 1969. The station was known as Cleckheaton Central from 2 June 1924 until it reverted to its original name on 12 June 1961. The station has one particular claim to fame: it is the only British station to have been stolen! A man was prosecuted – but found not guilty on the grounds that he had been duped – at Wakefield Crown Court in 1972 for stealing the material that resulted from the demolition of the station the previous year. The line through Cleckheaton to Thornhill remained open for freight traffic until the early 1980s – even surviving temporary closure caused by the construction of new bridges over the A58 and M62/M606 between 3 October 1970 and 1 April 1974 – and it was the route's final closure that was one factor in the development of the West Yorkshire Transport Museum project. Unable to fund the line's reopening to passenger traffic, West Yorkshire Metropolitan County Council sought to retain the route through another means. The creation of Transperience at Low Moor was designed ultimately to see the operation of preserved trams along the route; these proposals, however, came to nought and the line was eventually lifted. The trackbed survives intact throughout and now forms part of National Cycle Network route 66. *Neville Stead Collection/Transport Treasury*

Opposite above: **Pictured near** Lady Anne Crossing, Batley, on 6 August 1966, Class 5 No 45494, allocated to Crewe South shed, heads northbound with the 9.14am Saturdays Only service from Llandudno to Newcastle. *Mike Mitchell/Transport Treasury*

Opposite below: **Having just** departed from Dewsbury station, Class 47 No 47108 heads north with the 8.5am service from Liverpool to Newcastle on 3 July 1984. The eleven-arch Dewsbury viaduct, over which the train is passing, was designed by Thomas Grainger and completed in 1848; the structure is now Grade II listed. *Gavin Morrison*

THE HEAVY WOOLLEN DISTRICT • 109

On 18 May 2018, a Class 185 unit of TransPennine Express stands in Dewsbury station with a service towards Liverpool Lime Street. Following the award of the franchise to operate the TransPennine franchise to a joint venture between First Group and Keolis in 2003, the new franchisee undertook to replace the existing units used on the services with new three-car DMUs capable of speeds of 100mph. The contract to construct the fifty-six (late reduced to fifty-one) units was placed with Siemens and construction was undertaken at the company's factory at Krefeld in Germany and all were delivered during 2006 and 2007. All of the units were refurbished between June 2017 and 2018. Although the introduction of replacement rolling stock envisaged some of the Class 185s being released back to Eversholt Rail Group, in the event all remain in operation with TransPennine Express at the time of writing. Dewsbury station opened, courtesy of the Leeds, Dewsbury & Manchester Railway (albeit acquired by the LNWR before the line's opening), on 18 September 1848 and was modified by the LNWR in the late 1880s; the remaining structures are all Grade II listed. There were two other passenger stations in the town: Market Place, built by the LYR that closed on 1 December 1930, and Central, the GNR's station that closed on 7 September 1964. *Gavin Morrison*

LEEDS AREA

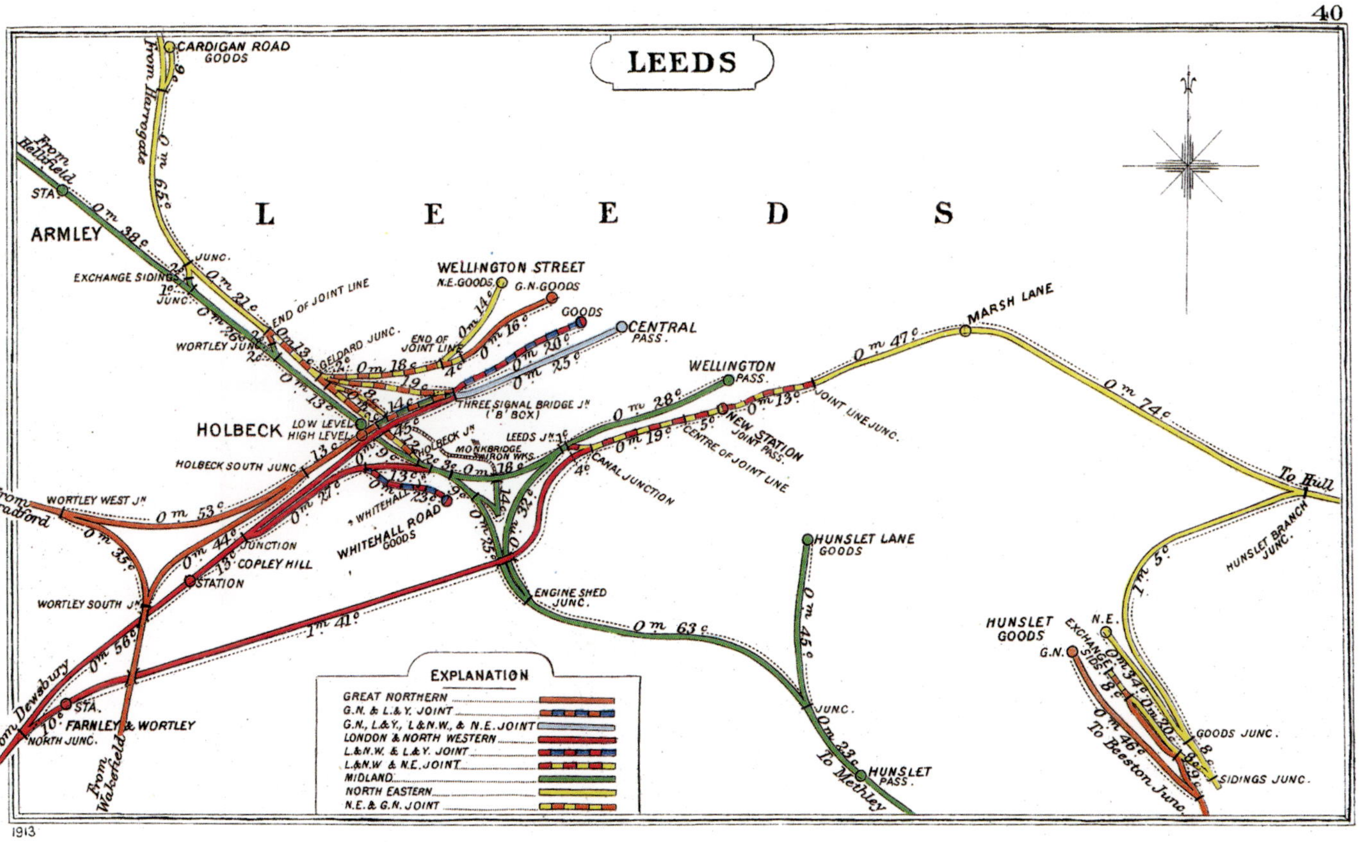

The Railway Clearing House map for Leeds dated 1913.

On 23 June 1957, two branches of the RCTS organised the 'Yorkshire Coast' rail tour that was scheduled to depart from Leeds City at 10.15am before heading to York. Amongst the lines visited were those to Easingwold, Kirbymoorside (by now the terminus of the route from Gilling to Pickering), Filey and Whitby. The rostered locomotive for the run from Leeds was Selby-allocated Class D20 4-4-0 No 62387, pictured here shortly prior to departure. By this date the locomotive was one of only a handful of the William Worsdell-designed class to remain in service. Completed at Gateshead Works in April 1907, the locomotive was reallocated to Tweedmouth shortly after the tour and was withdrawn in early September 1957. The last examples of the class were withdrawn later in the same year. *John McCann/Online Transport Archive*

Above: **The GNR's** 3¾-mile freight only branch from Beeston Junction to Hunslet, which opened on 3 July 1899, included a substantial viaduct that crossed over the reserved track section of Leeds Corporation's tram route to Middleton. Pictured in 1957 heading under the viaduct with an inbound service towards Cross Gates via the city centre is one of the corporation's ex-London Transport's 'Feltham' cars, No 525. The section of the railway branch from Beeston Junction over this viaduct to Parkside Junction was closed on 3 July 1967, the tramway having been closed on 28 March 1959. This area is now unrecognisable, having undergone major redevelopment. The section of the ex-GNR line from Parkside Junction to Hunslet closed on 3 January 1966; much the trackbed of this section was subsequently used for the route of the M621. *Phil Tatt/Online Transport Archive*

Opposite above: **On 18 June** 1958 Class 2P 4-4-0 No 40690 stands light engine outside Leeds City station adjacent to the ex-MR Leeds City Wellington signalbox. At the time, the locomotive, completed at Crewe Works in November 1932, was allocated to Holbeck shed, having been transferred there from Farnley Junction in October 1955. By this date the 4-4-0 was approaching the end of its operational life; it was withdrawn from Holbeck in October 1960. The box closed in 1967 with the rationalisation of the station and the completion of the resignalling scheme. The new power box at Leeds was commissioned between 29 April and 1 May 1967 and resulted in the closure of ten existing manual boxes (Leeds City West, Leeds City East, Leeds City Junction, Whitehall Junction, Wortley North, Leeds Central A, Leeds Central B, Holbeck Junction and Wortley South). *Paul de Beer/Online Transport Archive*

Opposite below: **On the** same day, Fairburn-designed 2-6-4T No 42072 stands in one of the erstwhile Wellington platforms at Leeds City awaiting departure. The condition of the trainshed over the ex-NER section of the station visible in the background is already poor; in less than a decade this structure was to disappear with the station's first post-war modernisation. When recorded here, No 42072 was allocated to Manningham shed; when new in November 1950, the locomotive was one of the batch initially allocated to the Southern Region. Migrating to Gateshead in November 1954, its transfer to Hammerton Street in July 1957 resulted in it spending the last decade of its life in the West Riding; it was withdrawn from Low Moor in September 1967. Two of the class – Nos 42073 and 42085 – were preserved following withdrawal, each having been withdrawn from Normanton in September 1967. *Paul de Beer/Online Transport Archive*

LEEDS AREA • 113

Above: Class V2 2-6-2 No 60812, allocated at the time to Heaton (52B) shed, awaits departure from the stygian gloom of Leeds City station with a westbound service on 13 July 1958. The 'V2', which had been completed at Darlington Works in September 1939, was to spend its entire operational life allocated to sheds in the Newcastle area, being withdrawn from Gateshead shed in July 1964. *Paul de Beer/Online Transport Archive*

Opposite above: The crew of 'A3' Pacific No 60082 *Neil Gow* acknowledge the photographer as the Down 'Thames-Clyde' express approaches Wortley Junction having just passed through Holbeck Low Level station. The junction provided a link between the ex-MR route from Leeds City towards Shipley with the joint (Great Northern and North Eastern railways) line from Central and the NER route thence to Harrogate and the north. In addition to the passenger traffic, the junction also provided access to the two goods yards at Wellington Street – belonging to the GNR and NER – as well as a connection to the Monkbridge Iron Works, which was situated to the north of the running lines, slightly to the east of Holbeck station. *Neville Stead Collection/Transport Treasury*

Opposite below: With the High Level station visible on the viaduct above, three-cylinder 'Compound' 4-4-0 No 41196 has just passed through Holbeck Low Level station with a service towards Leeds City. The latter station – jointly controlled originally by the Midland and North Eastern railways – opened on 2 June 1862. Although unofficially known as 'Low Level' prior to nationalisation, it was not until 2 March 1951 that the designation became official. Holbeck station – both the high and low level platforms – was closed on 7 July 1958. *Neville Stead Collection/Transport Treasury*

LEEDS AREA • 115

Above: On 8 April 1959, Class A3 No 60084 *Trigo* approaches Headingley with the Down 'Queen of Scots' Pullman service on 8 April 1959. The coaching stock visible here was largely replaced in the early 1960s following the introduction of BR's Mark I Pullman coaching stock. *Peter D.T. Pescod/Transport Treasury*

Opposite above: **The origins** of Leeds Central station were complex but arose from the needs of a number of railway companies that emerged during the period of the Railway Mania to serve Leeds. During the mid-1840s, a number of plans were proposed but it was not until January 1849 that a final agreement was achieved that saw a reduced structure based upon work that had already commenced. Even after this date there were disagreements amongst the participating companies that led to delays and the first services – to a temporary station at Wellington Street (converted from a warehouse and subsequently used as a goods shed after closure to passenger traffic) – operated by the LNWR commenced on 18 September 1848. In 1851, ownership of the station became jointly controlled by the LNWR, the LYR and the Leeds & Thirsk – although only the LYR's trains made use of the station – and it was not until 1854 that the GNR became a part owner. GNR services, which had served a further temporary Wellington Street station from 14 May 1850, were transferred to Central on 1 August 1854. Pictured in the station with some of the stock for the 'Queen of Scots' Pullman on 8 May 1960 is Class A1 No 60123 *H.A. Ivatt*. The 'Queen of Scots' Pullman service had its origins as the 'Harrogate Pullman' when first introduced on 9 July 1923; it was provided with new rolling stock in 1928 and extended through to Glasgow, being renamed the 'Queen of Scots' at the same time. Withdrawn as a wartime measure in 1939, the service was restored in 1952 but was destined to be cut back to Harrogate in 1964 (with a portion for Bradford from Leeds Central), being renamed the 'White Rose' at the same time. *Gavin Morrison*

Opposite below: **In 1960**, Class A3 No 60080 *Dick Turpin* heads the Up 'Thames-Clyde' express passed Armley Canal Road No 2 signalbox. The train is about to pass under the flyover that carried the fast lines over the slow lines; this was designed to ensure that local traffic on the line from Leeds City to Bradford Forster Square was separated from the main line traffic on the route between Leeds and Carlisle. Although the box was known as Armley Canal Road No 2, the diagram within the box carried the much simpler name Armley Junction. The box originally opened on 12 December 1889 and was to survive through until closure, following the commissioning of the Leeds power signalling box, on 23 April 1967. *Neville Stead Collection/Transport Treasury*

LEEDS AREA • 117

Above: **Standing at** the east end of Leeds City station on 2 August 1960 with the Down 'North Briton' is Neville Hill-allocated Class A3 No 60084 *Trigo*. In the background, Class K3 2-6-0 No 61814 can be seen awaiting departure with a service towards Bridlington. Although the origins of the 'North Briton' lay in the railway competition in the first decade of the twentieth century and the service was known unofficially as the 'North Briton' almost from its inception, it was not until the 1949/50 timetable that it was officially named. It ran from Leeds via York and Edinburgh to Glasgow. Dieselisation in the early 1960s saw the service's schedules improved but the name was dropped in 1968 (although it was briefly revived between 1972 and 1978 for a service from King's Cross to Dundee and later Aberdeen). The 'A3' would be based in Leeds until December 1963 when a final transfer took it to Gateshead, from where it was withdrawn the following November. *Mike Mitchell/Transport Treasury*

Opposite above: **On 7 August** 1960, two Class 5s – Nos 45428 (later preserved and named after the former Bishop of Wakefield and noted railway photographer Eric Treacy) and 45339 – are seen passing through Morley Low station with an empty coaching stock working from York to Manchester. The station, which was known as Morley from opening on 15 September 1848 through being renamed on 30 September 1951, was to lose the 'Low' suffix in the late 1960s following the closure of Morley Top station, on the ex-GNR route, earlier in the decade. *Mike Mitchell/Transport Treasury*

Opposite below: **With the** original main line to Huddersfield passing beneath, Class 5 No 45372 makes use of the flyover at Farnley Junction with a service towards Huddersfield via the 'Leeds New Line' on 30 June 1961. In the late nineteenth century, as traffic grew, the LNWR undertook improvements to the route between Huddersfield and Manchester in order to increase capacity; however, its main line between Huddersfield and Leeds require the use of a section of the LYR Calder Valley route between Heaton Lodge Junction and Dewsbury Junction. In order to alleviate this and to improve services between Leeds and Huddersfield through the Spen Valley, the 'Leeds New Line' was conceived. Authorised by two Acts – of 27 June 1892 and 6 July 1895 – freight traffic commenced on the section from Spen Valley Junction to Northorpe on 18 September 1899 and thence to Farnley Junction on 9 July 1900; passenger services were introduced on 1 October 1900. Although local passenger services were withdrawn in 1953, the section remained used for main-line services and as a useful diversionary route until closure in the mid-1960s. *Gavin Morrison*

LEEDS AREA • 119

Above: **On 9 July** 1961, the RCTS organised the 'Border' rail tour from Leeds via the Settle & Carlisle line to Carlisle and then over the Waverley route (with branches) to St Boswells, returning to Leeds via Kelso, Tweedmouth and the East Coast main line. Traction for the section from Leeds to Carlisle was provided by Stanier Pacific No 46247 *City of Liverpool*, which had been transferred from Camden (London) to Carlisle Kingmoor the previous month. The train, seen here at Leeds City, was scheduled to depart from the station on its outbound journey at 9.50am with a planned arrival back to Leeds at 9.20pm after more than 400 miles of travel. Leeds City station was created on 2 May 1938 through the merger of Leeds Wellington (opened as a permanent station on 1 October 1850) with Leeds New (opened on 1 April 1869). The train shed built for Leeds New – under which the train is pictured – was swept away during the rebuilding of the 1960s. *John McCann/Online Transport Archive*

Opposite above: **The station** at Beeston opened in February 1860, nearly three years after the opening of the Bradford, Wakefield & Leeds Railway route from Wakefield to Leeds on 5 October 1857. Seen passing the station on 16 August 1962 is Class B1 No 61115 with the 3.42pm service from Leeds Central to Doncaster. Beeston station had, by that date, already closed to passenger services – on 2 March 1953 – but looks to be substantially intact. Freight facilities continued to be provided at the station until 2 November 1964. In the sixty years since the station closed, all traces that it once existed have been eliminated although the main line – now electrified – continues in operation. *Mike Mitchell/Transport Treasury*

Opposite below: The station at Headingley opened on 9 July 1849 with the completion of the Leeds & Thirsk Railway's link between Leeds and Weeton via Bramhope Tunnel. Although two of the railway's directors – Henry Cowper Marshall and Christopher Beckett – lived in Headingley, their personal interests outweighed that of the railway company with the result that the station was located about a mile from the village itself. During the course of the nineteenth century, however, as Headingley expanded, the fields that surrounded the station were gradually fitted with new houses. The station building, which is now Grade II listed and viewed looking towards Horsforth in this photograph of 6 June 1964 (some five years after the building and the original platform were sold), was probably designed by Thomas Grainger, who was the Leeds & Thirsk's engineer. In the late nineteenth century, the need for higher platforms resulted in the construction of a new platform to the west of the original on the down line – the difference in height between the original and new is clearly visible in this view – and resulted in the staggered platforms that remain in use. *Neville Stead Collection/Transport Treasury*

LEEDS AREA • 121

Above: **On 1 August** 1964, Class 5 No 45352 approaches Farnley Junction over the 'Leeds New' line with the 9.30am Saturdays Only service from Manchester to Newcastle. The Up line in the foreground diverged from the original London to Huddersfield main line on the east and crossed over via a flyover. *Mike Mitchell/Transport Treasury*

Opposite above: **Heading northbound** with the 12.55pm Saturdays Only service from Blackpool to Leeds on 8 August 1964 is 'Jubilee' class 4-6-0 No 45643 *Rodney*. The train has just passed over A643 at Churwell heading towards Farnley Junction. There was a station at Churwell – located to the south of the viaduct over which the train has just passed – but this was closed completely on 2 December 1940 The new station at Cottingley – opened on 25 April 1988 – is sited slightly to the north of the view here. *Mike Mitchell/Transport Treasury*

Opposite below: **With the** Down 'Leeds New' line coming in from the south in the background, 'Jubilee' No 45562 *Alberta* is seen on 3 July 1965 having just passed Farnley Junction with the 9.3am Saturdays Only service from Leeds to Llandudno. At the end of its operational life, No 45562 was to achieve fame as one of the last two of the class to remain in service. At the start of 1967, the once 189-strong class had been reduced to only eight in operation; of these one was based at Low Moor, two at Wakefield and the remaining five – including No 45562 – at Holbeck. With the withdrawal of six of the survivors by the end of October only two of the class – Nos 45675 *Hardy* and 45562 – were still available for service; both of these were withdrawn during November 1967. Of the eight, one – No 45593 *Kolhapur* – survives in preservation. *Mike Mitchell/Transport Treasury*

LEEDS AREA • 123

Above: **On 16 July** 1965, Class B1 No 61337 pilots 'Britannia' class Pacific No 70010 *Owen Glendower* just south of Farnley Junction with the afternoon York to Manchester newspaper vans service. The locomotives have just passed under the single-track bridge that carried the 'Leeds New' line Up track from the junction towards the Spen Valley. The 'Britannia', which was completed at Crewe Works in May 1951 and initially allocated to Norwich Thorpe, was by July 1965 allocated to Carlisle Kingmoor, from where it would be withdrawn in September 1967. The 'B1' was allocated to York North from November 1948 until a final transfer, in June 1967, saw it move to Low Moor, where it was to spend the last five months of its operational life, before being withdrawn in the following November. *Mike Mitchell/Transport Treasury*

Opposite above: **Pictured approaching** Wortley South Junction on 28 August 1965 is Wakefield-allocated Class B1 No 61022 *Sassaby* with the 1.35pm Saturdays Only service from King's Cross to Bradford Exchange. The train is about to take the Wortley West curve, which provided direct access from the Wakefield to Leeds line to that from Leeds to Bradford. The curve survived until closure in 1985; its demise, however, sparked controversy in Bradford as it effectively removed the last surviving direct link from Bradford to London that bypassed Leeds. Such was the local feeling that City of Bradford MDC contemplated legal action against BR over the closure, although this was not pursued. *Mike Mitchell/Transport Treasury*

Opposite below: **Pictured approaching** the disused platforms at Holbeck High Level – the station had closed on 7 July 1958 – in 1966 is Class 5 No 44857 with a freight. The high level platforms at Holbeck had opened on 1 July 1855. Following the opening of the low level platforms seven years later, the station was known in certain publications – such as railway timetables – as Holbeck Junction although the use of 'Low Level' and 'High Level' was used on some tickets for a period the before the suffixes became officially recognised in March 1951. *Neville Stead Collection/Transport Treasury*

LEEDS AREA • 125

Above: **It is** 6 August 1966 and now just over a month since the station at Armley Moor had closed (on 4 July) as 2-6-4T No 42184 passes through with the 5.10pm service from Bradford Exchange to King's Cross. Although the line itself remains operational, all trace of the station – including the platforms – has now disappeared with the land between Station Way and the running lines having been redeveloped. *Mike Mitchell/Transport Treasury*

Opposite above: **Opened with** the line on 1 August 1854, Armley Moor station was to lose its passenger services on 4 July 1966. A month later – on 6 August – the photographer makes use of the closed platforms to record Class B1 No 61030 *Nyala* heading west with the 1.40pm service from King's Cross to Bradford Exchange as the train tops the 1 in 50 gradient on the approach to the station. Originally known as Armley & Wortley when first opened, the station was renamed on 25 September 1950. *Mike Mitchell/Transport Treasury*

Opposite below: **Class 5 No 44694** is pictured approaching Bramley station with a westbound service towards Bradford Exchange on 10 September 1966. The two-platform station opened, courtesy of the Leeds, Bradford & Halifax Junction Railway on 1 August 1854, and had closed to passenger traffic just over two months prior to the date of this photograph, on 4 July 1966. A new station, with staggered platforms straddling Skinnow Road (the road under the bridge in the foreground), was opened on 12 September 1983. The junction for the route via Pudsey Lowtown towards Laisterdyke was situated slightly to the west of Bramley station. By 1966, the Pudsey loop had closed completely. *Neville Stead Collection/Transport Treasury*

LEEDS AREA • 127

The LNER timetable for the period from 16 June to 5 October 1947 covering the lines from Bradford Exchange to Leeds via either Pudsey or Stanningley and for the services between Leeds and Halifax via Low Moor.

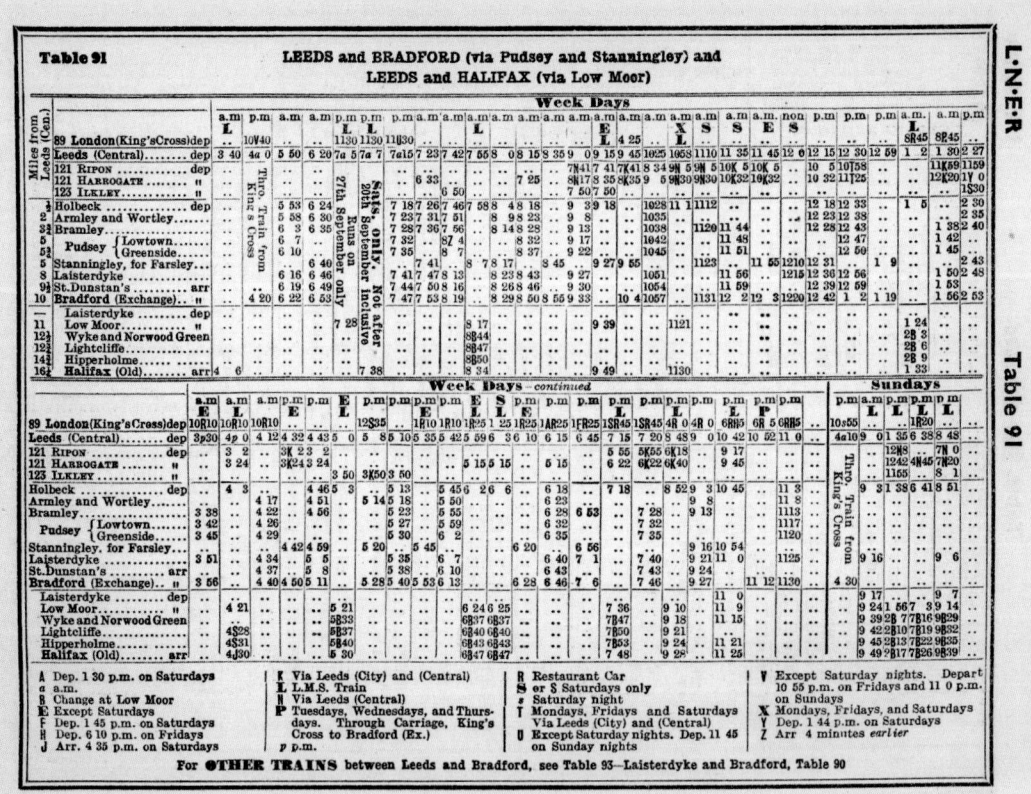

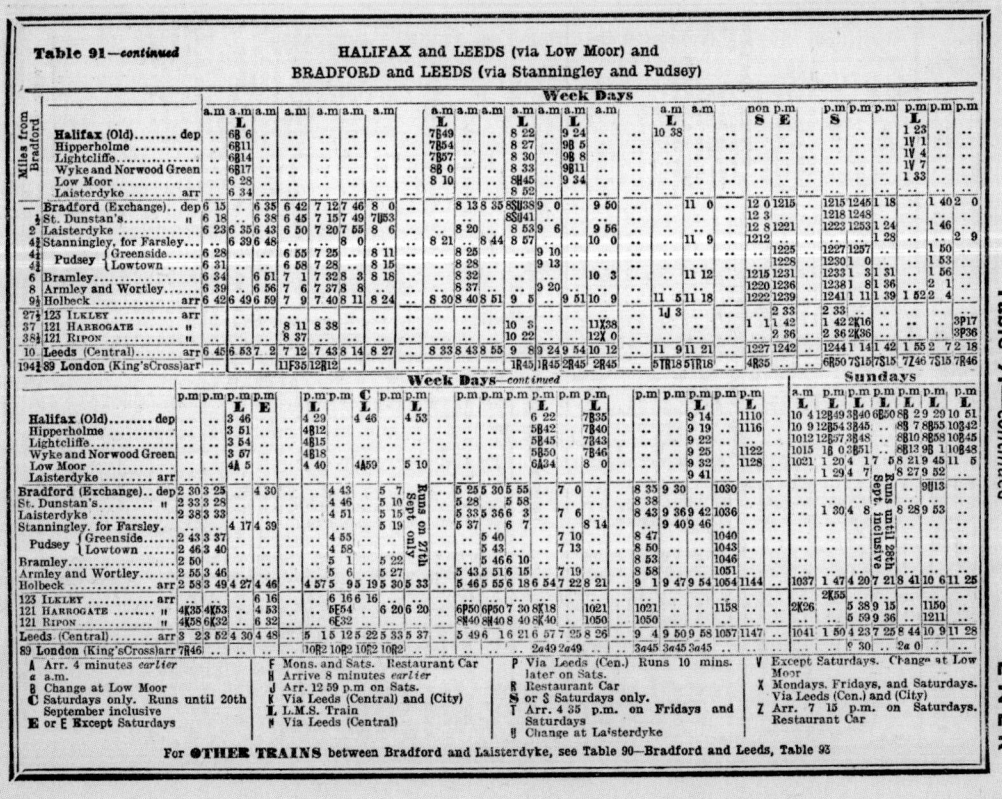

On 18 February 1967, towards the end of the station's life, Fairburn-designed 2-6-4T is seen awaiting departure from Leeds Central station. The reduction in services during the early 1960s, allied to the planned withdrawal of local services to Skipton via Ilkley and Bradford via Shipley, enabled the decision to be made to concentrate all passenger services on Leeds City once a new curve at Whitehall Junction to enable services over the ex-GNR line from Wakefield to gain access to the ex-MR line through Holbeck into City station. Central station was closed on 1 May 1967 – the last train from the station having operated on 29 April – and was subsequently demolished and the site redeveloped. The viaduct that carried the line from Holbeck to Three Signal Bridge Junction over the River Aire and Leeds-Liverpool Canal, however, remains intact as does the tower that contained one of the goods lifts. *Gavin Morrison*

The first new station to be opened in the area for more than a generation was New Pudsey. It was opened in conjunction with a number of road improvements in the area, including the construction of the A647 (the road overbridge for the new road, still being completed, is visible as a backdrop in this view). On 18 March 1967 – less than a fortnight after the station opened on 6 March – 2-6-4T No 42196 is seen arriving in the newly-built platform with an Up service from Bradford Exchange. The station, with its large car park, was one of the earliest of the stations to be constructed by BR to cater for car-borne traffic. *Gavin Morrison*

Above: **On 17 June** 1967, Fairburn-designed 2-6-4T No 42073 – one of the two of the class later to be preserved – approaches the old station at Stanningley with the Down 'Devonian' towards Bradford Exchange. The 'Devonian' was launched by the LMS in 1927 and covered a through service from Bradford Forster Square to Bristol and, courtesy of the GWR, to Paignton. The name was suspended during the Second World War but revived in October 1946. As part of the planned run-down of Forster Square station, the 'Devonian' was transferred from 1 May 1967 to operate from Bradford Exchange. The name was again dropped in 1975 but was revived in part of the north-east/south-west HST timetable in 1982. The final Bradford link to the service ceased with the start of the May 1987 timetable when the southbound departure was cut back to Leeds. The name itself was finally discontinued in 2002. *Gavin Morrison*

Opposite above: **As a** member of the locomotive crew keeps a wary eye on the photographer, 2-6-4T No 42138 heads the 7.45am Saturdays Only service from Bradford Exchange to Great Yarmouth between Wortley Junction and Beeston on 8 July 1967. By this date steam working in the Bradford and Leeds area was becoming increasingly uncommon as the process of dieselisation continued apace. No 42138 had been based in the area since it had been transferred to Manningham from Kentish Town in London during November 1954. When recorded here it was allocated to Normanton but was approaching the end of its seventeen-year operational life as it was withdrawn only two months after the date of this photograph, in early September 1967. *Mike Mitchell/Transport Treasury*

Opposite below: **Pictured awaiting** departure from the newly rebuilt Leeds City station on 28 July 1966 with the 10.15am service to Bradford Forster Square is 2-6-4T No 42138. The train had originally departed from Birmingham New Street at 6.40am. Although there had been plans for the development of a single station serving Leeds for almost a century, it was not until 1959 that the British Transport Commission developed a scheme, costing, £4½ million, for the upgrading of City station. However, cutbacks in the early 1960s resulted in the project being reduced in scope. The work resulted in the elimination of most of the terminal platforms from the erstwhile Wellington part of the station; these were either filled in for use as a car park or were retained solely for parcels traffic. The new overall roof – which was designed to accommodate 25kV if it ever reached Leeds – was completed in May 1966 and was thus almost new when recorded here. With the closure of Central on 29 April 1967, the long-term aspirations of having a sole station were finally realised. *Mike Mitchell/Transport Treasury*

LEEDS AREA • 131

Above: **A contrast** in front ends at Leeds City on 8 September 1969 sees Type 2 (later Class 24) No 5113 alongside Type 4 (later Class 45) No 53 *Royal Tank Regiment*. The backdrop shows to good effect the profile of the new station built in the mid-1960s; the metal canopy, completed in 1967, was to survive until the most recent rebuilding of the station, undertaken between 1999 and 2002. *Neil Caplan/Online Transport Archive*

Opposite above: **On 13 September** 1975, a four-car DMU is pictured heading eastbound through Marsh Lane cutting as it runs parallel with East Park Road towards Neville Hill depot. The line through the cutting formed part of the route of the Leeds & Selby Railway, which opened from a terminus at Marsh Lane through to Selby on 22 September 1834 when a single line was completed; the second line was opened by 15 December 1834. It was only from the later date that the railway commenced carrying freight traffic. The new railway reached its Marsh Lane terminus through a 700-yard-long tunnel under Richmond Hill. The Leeds & Selby became part of the York & North Midland in 1844 and thus became part of the NER. It was under the NER that the network of lines to the east of Leeds expanded with the result that the existing two-tracks through Richmond Hill proved inadequate and in the late nineteenth century work commenced on the widening of the route. The tunnel was eliminated and Marsh Lane cutting opened up; the work on the widening was completed in 1895. *Chris Gammell/Online Transport Archive*

Opposite below: **A three-car** DMU heads westwards at Armley Moor amidst evidence of recent track work. With the church of St Bartholomew dominating the background, the DMU is passing the site of the long-closed station on 1 August 1977 heading towards Bradford Interchange. The coal sidings were accessed via the spur towards the right and the track was modified towards the end of the sidings' life; coal traffic to the yard ceased in the early 1980s. The scene today is radically different; all surviving traces of the station have been removed and its site, plus that of the foundry in the background demolished with the whole area redeveloped. The sidings have been eliminated and the site again redeveloped. There are only effectively three constants: the church in the background, the road overbridge and the running tracks between Leeds and Bradford. *Gavin Morrison*

Above: **Seen passing** Pepper Road, Hunslet, on 18 October 1979 is Class 40 No 40165 with the 09.35 service from Carlisle to Nottingham. In the distance an '08' shunter can be seen shunting the Balm Road sidings. The junction for the link through to the – by now – preserved Middleton Railway is situated at the northern end of Hunslet Sidings. Completed at Vulcan Foundry in October 1961, No 40165, which was allocated to Edinburgh Haymarket for its entire operational life, was approaching the end of its twenty-year career when recorded here; it was withdrawn in July 1981. *Gavin Morrison*

Opposite above: **Allocated to** Finsbury Park for its entire operational career, Class 55 No 55015 *Tulyar* was approaching the end of its twenty-year life on the main line when recorded at Leeds City with an Up service towards King's Cross on 2 April 1981. Like all of the class, No 55015 had been built by English Electric at Vulcan Foundry, entering service in October 1961. Withdrawn in early January 1982, *Tulyar* is one of the type to survive in preservation. *Gavin Morrison*

Opposite below: **On 14 April** 1981, a five-car DMU – headed by Class 104 No E50549 – approaches Leeds City station from the west. By this date, BR's first generation DMUs were starting to reach the end of their long operational life; No E50549, allocated to Neville Hill (although spending some time based at South Gosforth on Tyneside), was some twenty-six years old and was destined to be withdrawn some six months after the date of this photograph. *Geoffrey Tribe/Online Transport Archive*

LEEDS AREA • 135

Above: **On the** same date, Class 08 No 08245 at the head of a short train of oil tanks is also seen approaching Leeds City; completed at Darlington Works in July 1957 as No 13315 and renumbered D3315 in June 1963, No 08245 was allocated to York (North) for the bulk of its career but was to end its days at Neville Hill, from where it was withdrawn in May 1984. Class 08 shunters were a familiar sight at Leeds City, often being employed on the shunting of parcels stock in the former Wellington platforms until the disappearance of this type of traffic and the rebuilding of the station. *Geoffrey Tribe/Online Transport Archive*

Opposite above: **Both the** GNR and NER constructed freight-only branches to serve Hunslet; the former was opened from Beeston Junction on 3 July 1899 whilst the latter, from Neville Hill West Junction, opened on 2 January 1899. Whilst the former GNER route was closed in two stages – from Parkside Junction to Hunslet on 5 January 1966 and Parkside Junction to Beeston Junction on 3 July 1967, the ex-NER route survives, despite the closure of the original NER goods yard on 5 September 1966, to serve a number of stone terminals. One of the sidings also served by the ex-NER branch was the Shell oil terminal at Hunslet and, on 4 April 1984, Class 40 No 40056 is pictured at the terminal with a service from the refinery at Stanlow. The site of the oil depot, which closed during the latter half of the first decade of the twenty-first century, has now been fully redeveloped. By the date of this photograph, No 40056, one of a class that was familiar in the area for many years, was approaching the end of its career; it was withdrawn in September 1984 and scrapped in early 1985. *Gavin Morrison*

Opposite below: **With parcels** stock standing in the background, Class 141 Pacer DMU No 141017 is seen in Leeds City station on 21 March 1987 in its original West Yorkshire PTE livery of Verona green and buttermilk. The twenty-strong class, based upon the bus body designed for the Leyland National, was constructed by British Leyland. Based at Neville Hill, the units, Nos 141001-20, proved troublesome in operation and during 1988 and 1989 all twenty underwent modification at Hunslet-Barclay, being renumbered Nos 141102-20/01 respectively at the same time. They were also repainted into the later West Yorkshire livery of red and cream. No 141017 – in its guise as No 141118 – was modified during the 1990s by SERCO for use as a weedkilling train. Following withdrawal, twelve of the class, including No 141118, were exported to Iran and two to The Netherlands. Of the remainder, No 141104 was scrapped, following an accident on 6 November 1989. Also scrapped was No 141101 but initially four – Nos 141103/108/110/113 – were preserved but two of these – Nos 141103 and 141110 – were subsequently scrapped as well. *Geoffrey Tribe/Online Transport Archive*

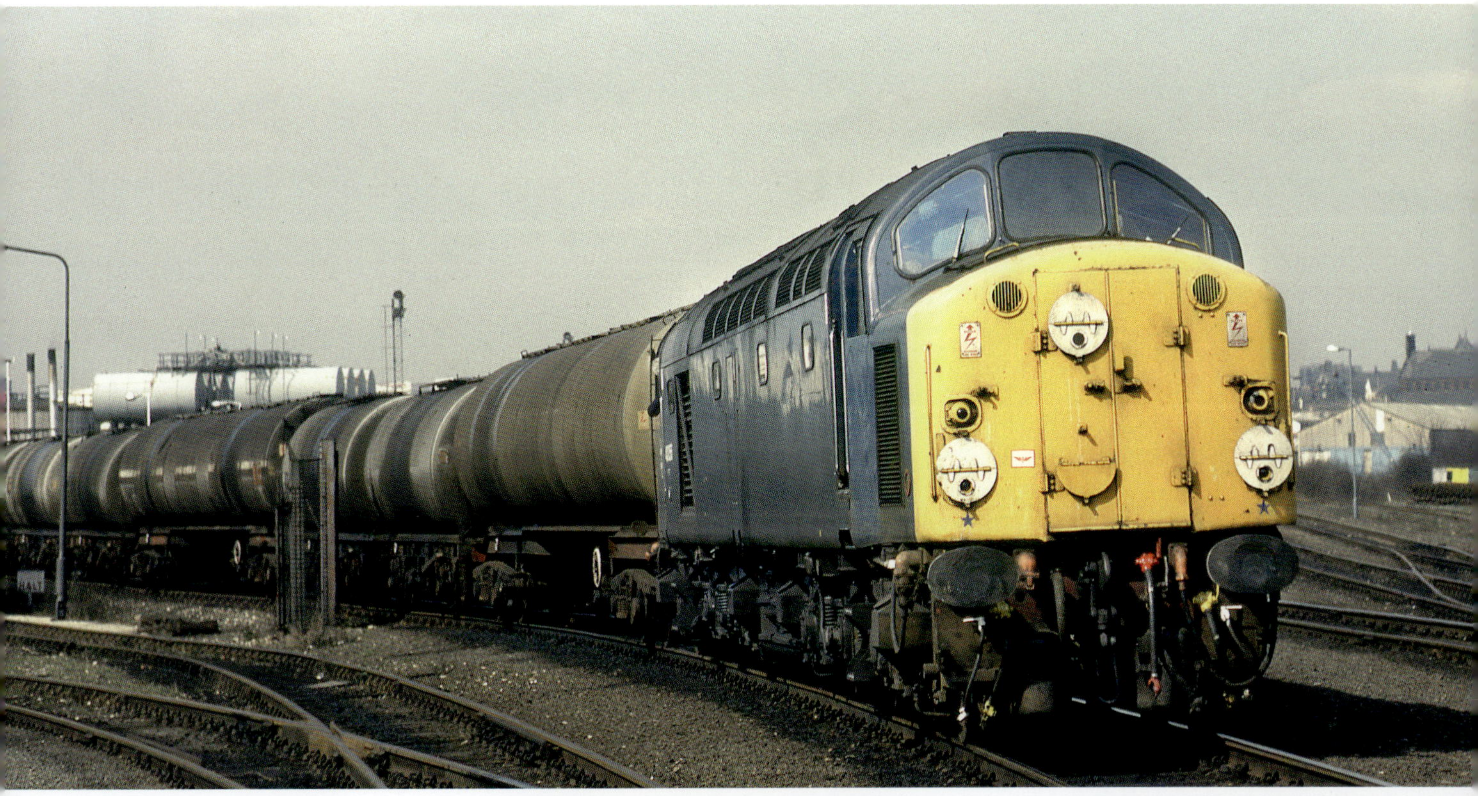

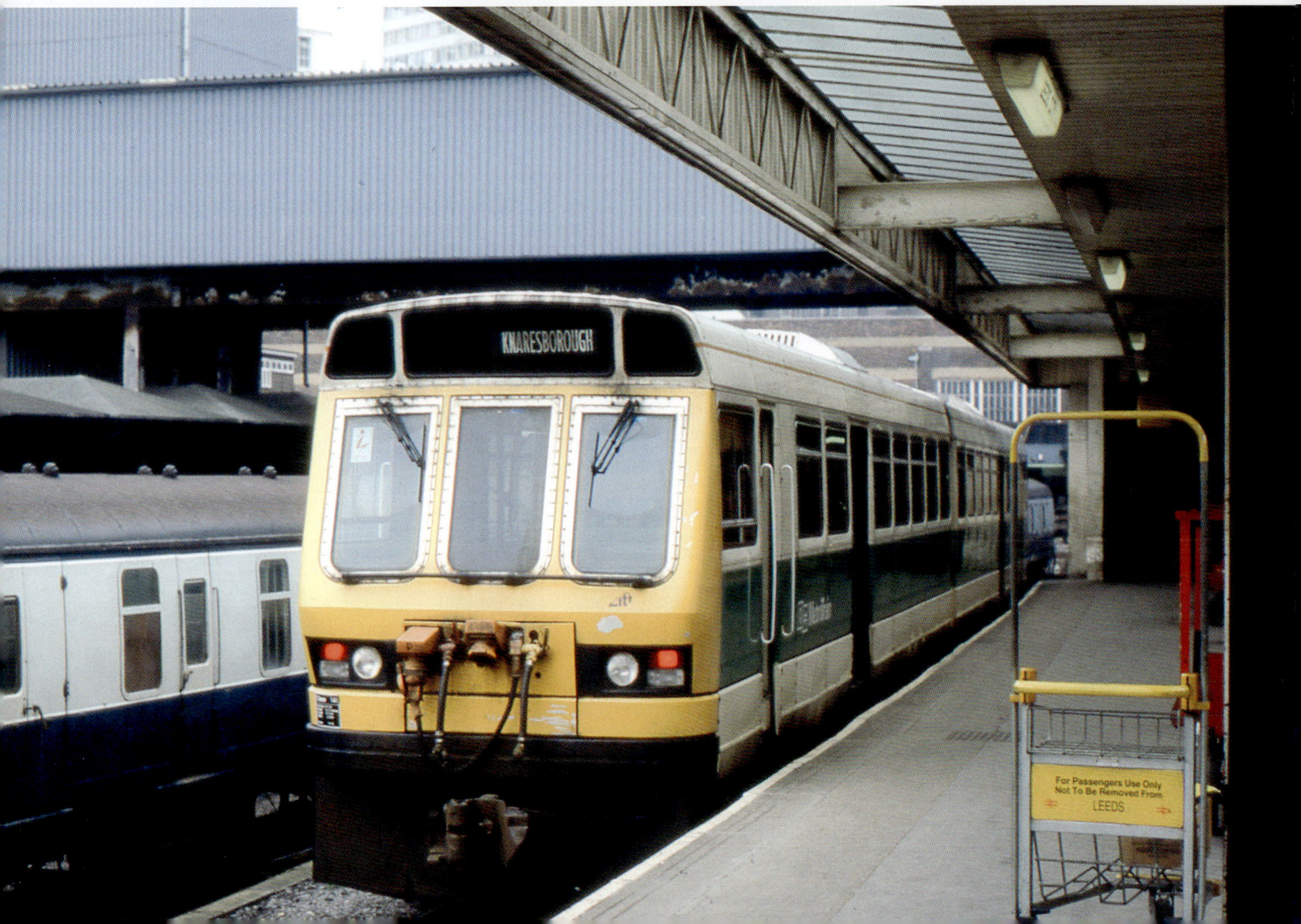
KNARESBOROUGH
For Passengers Use Only
Not To Be Removed From
LEEDS

For the introduction of electrified services over the East Coast main line to Leeds and Edinburgh BR introduced the thirty-one-strong Class 91. The locomotives, which were built at Crewe between April 1988 and March 1991, were designed to operate with Mk 4 coaching stock and a Driving Van Trailer; branded InterCity 225, operation commenced in 1989. Following privatisation, the original franchisee for the East Coast main line section was Great North Eastern Railway, which commenced operation on 28 April 1996. Here No 91028 is seen entering Leeds with the 11.37 service from Bradford Forster Square to King's Cross on 24 February 1999. Over the past quarter-century, the East Coast franchise has proved to be one of the more difficult ones to maintain; following GNER, which lost the franchise in 2007, there have been four further operators of the route – National Express East Coast (2007-09), East Coast (2009-15), Virgin Trains East Coast (2015-18) and London North Eastern Railway (since 2018) – two of which were (or are) subsidiaries of Directly Operated Railways (the Department for Transport company established to be the operator of last resort). The introduction of the new 'Azuma' units was to see the start of Class 91 withdrawals although, at the time of writing, LNER retained a number of sets in service. No 91028 (later renumber 91128) is one of the batch now taken out of service and stored; it has been used subsequently as part of a test train on the Midland main line. *Gavin Morrison*

When the Class 333 EMUs were ordered by Angel Trains in 1998 for use on the Airedale and Wharfedale lines, the original specification was for sixteen three-car sets. These were constructed by CAF in Spain with first deliveries being received at Neville Hill in March 2000. The first of the units entered service on 12 January 2001; by this date, however, it had been decided to order a further eight trailer coaches to convert the first eight units in four-car sets and it is in this condition that No 333008 is pictured at Leeds in 2003. A further eight trailer coaches were purchased with all sixteen being converted into four-car sets by the end of 2003. *Geoffrey Tribe/Online Transport Archive*

On 21 May 2020, three of the new Class 68 diesel-electrics leased by TransPennine Express – Nos 68029/028/032 – pass through Morley station light engine as they run from Crewe to York. A total of thirty-four of the type were ordered by Direct Rail Services and delivered between 2013 and 2017. They were manufactured in Spain initially by Vossloh España and then by its successor Stadler Rail. Of these, fourteen, later increased to sixteen, were leased to TransPennine Express for use with Mark 5A coaches. The locomotives – Nos 68019-034 – and coaches are destined for use on the services from Liverpool Lime Street to Scarborough and from Manchester Airport to Redcar. *Gavin Morrison*

When constructed, New Pudsey station was provided with platforms capable of accommodating InterCity services; however, the electrification of the Leeds to Bradford Forster Square line resulted in the transfer of the through King's Cross to Bradford services to the ex-MR route and the line from Leeds via New Pudsey to Interchange was relegated to local services only. On 30 June 2010, Class 158 No 158901 is pictured heading eastbound at the station. *Gavin Morrison*

On 16 April 2020, GBRF Class 66 No 66779 – the last of the type to be constructed – is seen passing Pepper Road, Hunslet, with the 11.25 service from Acrow Quarry to Pendleton. Since the view on page 134 of No 40165 passing this location, the former Balm Road Sidings have been replaced by Freightliner's Leeds Midland Road maintenance depot. The depot was opened by Pete Waterman's London & North Western Railway Co Ltd in July 2003 as part of a ten-year contract to maintain Freightliner's Class 66 locomotives. The site was purchased by Freightliner Maintenance Ltd in 2006 and it currently handles both Classes 66 and 70 as well as wagon repairs. The line of the proposed HS2 route to Leeds – the abandonment of which was announced as this book was being completed – made use of the alignment on which the depot stands. *Gavin Morrison*

THE NORTH EASTERN LINES FROM LEEDS

There was one intermediate station on the NER line between Otley and Arthington; this was Pool-in-Wharfedale, viewed here from the east in 1957. Although there had been proposals for the construction of a line from the Leeds to Thirsk main line west towards Otley and beyond in the 1840s, it was not until two decades later that the line opened. Construction began in 1863 with the first service from Leeds to Otley running on 1 February 1865. Pool station generated a considerable amount of freight traffic; the coal drops visible on the north side of the station supplied the local coal merchant, latterly Robert Midgley & Son, whilst the stone building visible on the left was the Pool Bank quarry stone dressing shed. A connection to the Pool Bank quarries owned by Benjamin Walker & Son was opened in 1880. Additional traffic was also generated by Whiteleys Paper Mill. Passenger traffic over the line ceased on 22 March 1965, with freight facilities being withdrawn from the station on the following 5 July when the line closed completely. The track was lifted the following year and the site of the station was redeveloped for housing eight years later. The trackbed west from Pool is currently under conversion into a cycleway. *Neville Stead Collection/Transport Treasury*

Above: The first station south of Wetherby on the line to Crossgates was Collingham Bridge and on 23 April 1957, Class J39 No 64920 is seen passing through the station with a return working race special from Wetherby Racecourse station. When the station opened on 1 May 1876, it was provided with a single platform as the line was single track until it was doubled in 1901. A second platform was completed on the west side; this was equipped with a wooden waiting shelter. *Mike Mitchell/Transport Treasury*

Opposite above: Amongst the lines visited by the RCTS's 'Roses Rail Tour' of 8 June 1958 was the network that served the Royal Ordnance Factory at Thorp Arch, served by the Wetherby to Church Fenton line. With the ever-increasing likelihood of war with Germany after the Nazi Party came to power in 1933, re-armament became a more urgent requirement. A number of ROFs were constructed from the late 1930s onwards; one of these was situated at Thorp Arch where work on the site commenced in early 1940. In all, the ROF eventually occupied a site of 642 acres with a railway connection first being laid on 24 June 1940. Initially, workers to and from the ROC were handled at the small LNER station at Thorp Arch but this proved inadequate, despite being lengthened, and during 1941 and 1942 four platforms – River, Ranges, Roman Road and Walton – were constructed on a circular line that surrounded the site; this was completed on 19 April 1942. This view, taken from the DMU used on the special, shows Roman Road platform. By the date of the special, the ROF was approaching the end of its military career; production had ceased in April 1958 and the site was sold in March 1959. After the site was purchased for conversion into a trading estate, the railway was lifted. *John McCann/Online Transport Archive*

Opposite below: In order to control access into and out of the sidings associated with the ROF at Thorp Arch, the LNER constructed a new signalbox – Thorp Arch East – which is seen here from the lead cab of the DMU used on the RCTS's 'Roses Rail Tour' of 8 June 1958. The original signalbox that served Thorp Arch station and the western access to the ROF was renamed Thorp Arch West at the same time. *John McCann/Online Transport Archive*

THE NORTH EASTERN LINES FROM LEEDS • 143

Above: The timetable for the Leeds to Wetherby line via Thorner covering the period from 16 June to 5 October 1947 – the last summer of the LNER's existence. *Author's Collection*

Opposite above: Otley station viewed from the south-west in 1959 sees an 0-6-0 shunting freight wagons. The station opened on 1 February 1865; the section west towards Ilkley was owned by the Otley & Ilkley Joint, whilst that eastwards to Arthington belonged to the NER. In 1947 there were three services on weekdays only in each direction between Bradford and Harrogate with six from Leeds to Ilkley and five in the reverse direction; the Leeds to Ilkley section merited a Sunday service, albeit only of two trains in either direction. Passenger services were withdrawn on 22 March 1965 with freight traffic succumbing on 5 July 1965. Following closure, the station site was cleared although it is still possible to trace much of the trackbed. *Neville Stead Collection/Online Transport Archive*

Opposite below: On 28 December, Class J39 0-6-0 No 64922 is seen taking water at Wetherby with a parcels service from Ripon to Leeds. The original station serving Wetherby had opened on 10 August 1847 with the line from Church Fenton to Harrogate. However, the opening of the line from Crossgates on 1 May 1876 meant that services from Harrogate to Leeds via the line to Crossgates, introduced following the completion of the south to west curve in 1901, were unable to serve Wetherby. As a result, the station was relocated south of the triangular junction with the new facility (on Linton Road) being opened on 1 July 1902. The original station (on York Road) remained in use for freight traffic, although limited freight – horse boxes, cattle vans, etc – was also handled at the new station as evinced in this view from the south. Until 18 May 1959, Wetherby was also served by a racecourse station; this was not shown in timetables but was located slightly to the east of the original station. When this closed, passenger traffic to and from the racecourse was handled at the 1902 station, with buses being used to transport passengers to the course itself. The 1902 station closed completely on 6 January 1964 although freight traffic continued to the old station, served from the surviving section of the Church Fenton line until 4 April 1966. The William Bell-designed station of 1902 remained extant, albeit increasingly derelict, until the early 1970s and was then demolished. The site was subsequently levelled for use as a car park. *Mike Mitchell/Transport Treasury*

THE NORTH EASTERN LINES FROM LEEDS • 145

Above: **Another Class B16,** No 61414, along with 2-6-0 No 43111, is seen passing through Scholes station on 18 April 1960 with a race special from Sheffield to Wetherby. As with the other intermediate stations on the line, Scholes was originally provided with a single platform; the Down platform was added in 1901 with the doubling of the route. The photograph shows to good effect the wooden platform shelters constructed by the NER when the stations on the line was provided with second platforms. Unlike other station buildings on the route, that at Scholes still survives although it has been substantially rebuilt to form a pub and restaurant. No 61414 was another of the unrebuilt members of the class and was also allocated to Neville Hill when recorded here; it was reallocated to Mirfield in December 1960 from where it was withdrawn in September 1961. No 43111, new from Doncaster Works in July 1951, was allocated to Sheffield (Grimesthorpe) shed and had brought the service through from Sheffield; the 'B16' had been added at Leeds to provide additional power on the section through to Wetherby. *Mike Mitchell/Transport Treasury*

Opposite above: **The sidings** at Collingham Bridge were used for the stabling of the rolling stock used on the specials to Wetherby Racecourse and, on 19 April 1960, 2-6-0 No 42751 has just arrived from Wetherby with the empty coaching stock from a Bradford to Wetherby racecourse special. A further rake of coaches is already standing on the loop behind the down platform. This view shows to good effect the track leading to the goods yard and coal drops. The last of the Bradford to Wetherby racecourse specials operated in 1963, the year in which the lines from Crossgates to Wetherby and Church Fenton to Harrogate were both slated for closure in the Beeching report. Following a three-month enquiry, closure of the two routes was confirmed on 24 October 1963. The station was closed completely on 6 January 1964 and the station demolished in the early 1970s. The site has been subsequently developed for housing. *Mike Mitchell/Transport Treasury*

Opposite below: **On 19 April** 1960 Class B16 4-6-0 No 61415 is seen passing through Thorner station with a return racecourse special from Wetherby. The station, with its single platform originally, opened with the line on 1 May 1876 when it was known as Thorner & Scarcroft. The name of the station varied over the next two decades, but it became simply Thorner on 1 May 1901; the same year saw a second platform added with the doubling of the line. As can be seen in this view, the goods yard and shed were located south of the station on the Down side. Both passenger and freight traffic were withdrawn from 6 January 1964 and the site was redeveloped for housing. The 'B16' class was designed by Sir Vincent Raven, with seventy being constructed between 1919 and 1924. Although sixty-nine of the type passed to BR in 1948, a number – but not No 61415 – had been rebuilt to the designs of Sir Nigel Gresley (designated Class B16/2) and Edward Thompson (Class B16/3). No 61415 had originally been completed at Darlington Works in September 1920 and spent the last five years of its life, following a final reallocation, based at Neville Hill. It was withdrawn in September 1961. *Mike Mitchell/Transport Treasury*

THE NORTH EASTERN LINES FROM LEEDS • 147

Above: **On 12 August** 1961, Class A3 No 60072 *Sunstar* is pictured heading south between Weeton and Arthington with the 10am service from Sunderland to Manchester. The locomotive had been based at Holbeck shed from November 1960 but had been transferred to Heaton shed in Newcastle about a month before the date of this photograph. New in September 1924 as LNER No 2571, the locomotive had been converted into an 'A3' in July 1941. By August 1961, No 60072 was approaching the end of its career; it was finally withdrawn in October 1962. *Mike Mitchell/Transport Treasury*

Opposite above: **The station** at Arthington was originally opened by the Leeds & Thirsk Railway in July 1849 when it was known as 'Pool'; the name was changed to Arthington three years later. The station's original location was about a half-a-mile north of the station illustrated here to the south of Arthington Lane and adjacent to the Wharfedale Inn. The station was resited in conjunction with the opening of the line to Otley on 1 February 1865. As can be seen, platforms were provided on two sides of the triangle; the third side, without platforms, can just be discerned in the distance. The station closed on 22 March 1965 and the station was subsequently demolished, although elements of the gasworks, opened by the NER in 1876 within the triangle to supply gas for lighting on trains and at stations, survive more than a century after it closed (in 1905). *Transport Treasury*

Opposite below: **On 13 June** 1962, the East Midlands Branch of the RCTS organised its 'East Midland No 5' rail tour from Nottingham Victoria to Darlington and return. For its outward journey, the train travelled from Church Fenton via Wetherby (where it reversed), Harrogate and Ripon rather than travelling over the East Coast main line. The special was hauled by SR 'Schools' class 4-4-0 No 30925 *Cheltenham* (which was later preserved) and LMS 4-4-0 No 40646, thus bringing two unusual locomotive types to the ex-NER line from Church Fenton to Wetherby. The train, with No 30925 leading, is seen here at Newton Kyme just after midday as it makes its way towards Wetherby. The station, opened as Newton with the line on 10 August 1847, was renamed Newton Kyme in August 1850. The main station building was designed by G.T. Andrews. Passenger services ceased on 6 January with freight following six months later on 6 July. The line remained operational serving Wetherby until 4 April 1966. Following closure, the station building was converted into a private house. *Neville Stead Collection/Transport Treasury*

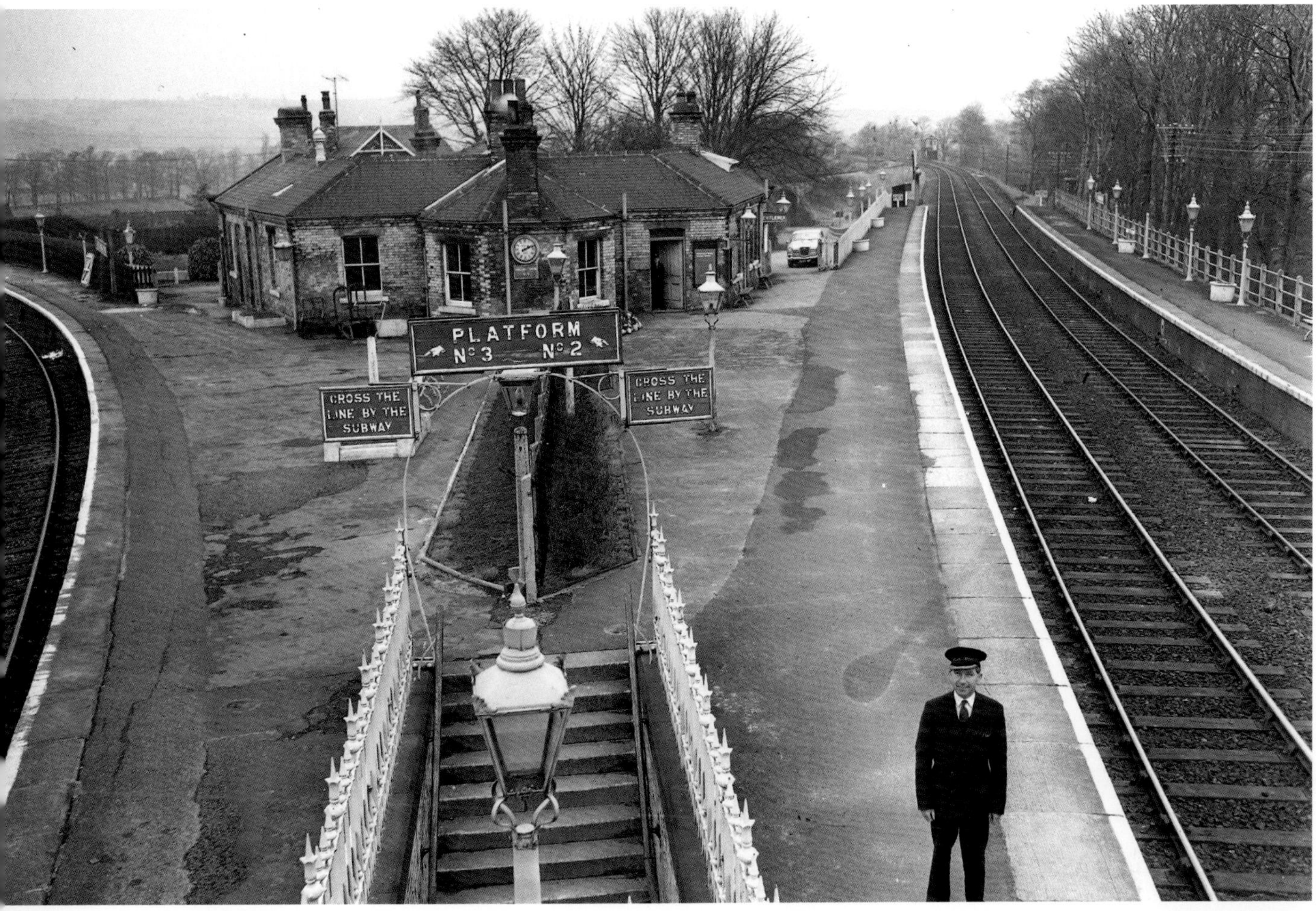
PLATFORM
Nº 3 Nº 2
CROSS THE LINE BY THE SUBWAY
CROSS THE LINE BY THE SUBWAY

R.C.T.S.
THE
EAST MIDLANDER
30925

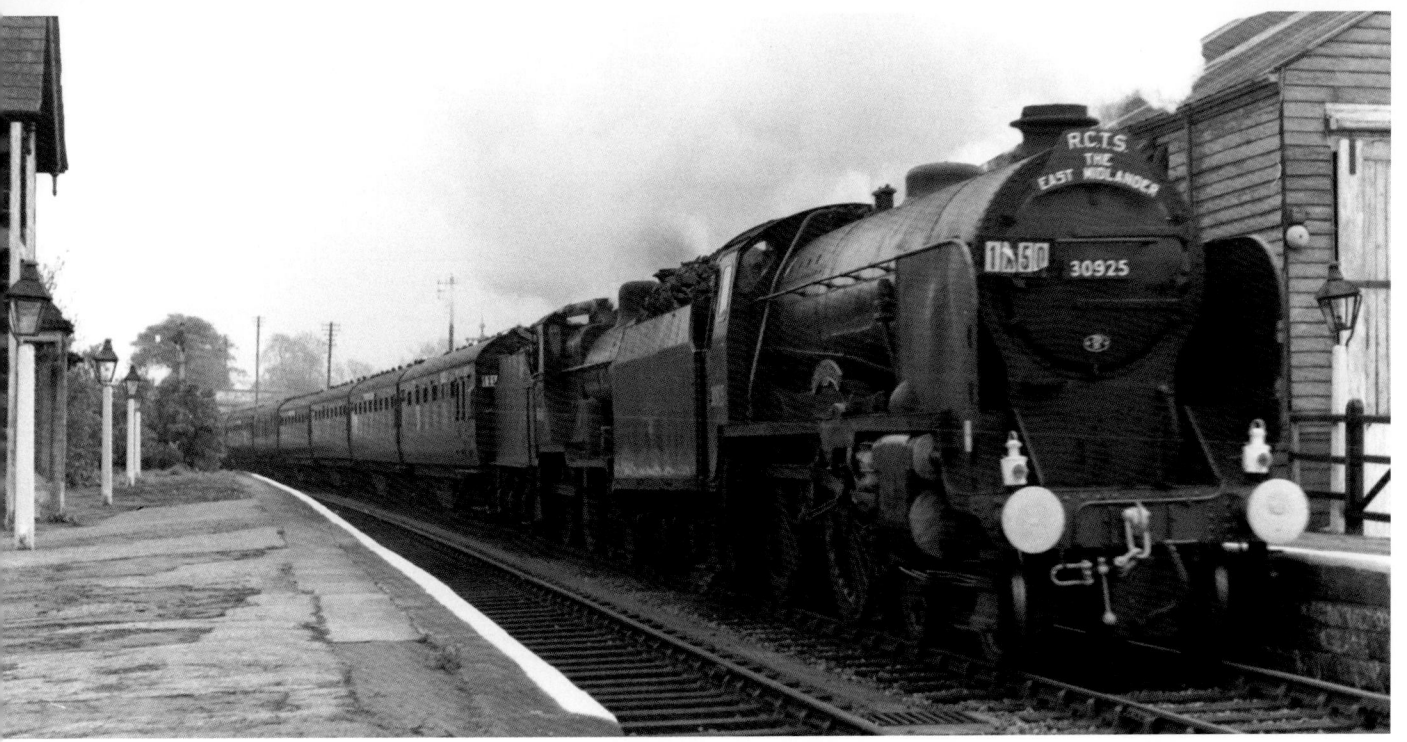

Above: **In 1963,** Class Q6 0-8-8 No 63445 heads west towards Leeds with a loaded coal train as it passes the junction with the line south through Kippax and Ledston to Castleford. The latter line had opened throughout to freight traffic on 8 April 1878 and to passenger traffic on the following 12 August. By the date of the photograph, passenger services over the line to Castleford had been withdrawn – on 22 January 1951 – but the line remained open for freight traffic until 14 July 1969 when the section between Garforth and the colliery at Ledston closed completely. *Neville Stead Collection/Transport Treasury*

Opposite above: **When the** line from Cross Gates to Wetherby opened on 1 May 1876, the junction at Wetherby faced towards Church Fenton, thus precluding a direct service from Leeds to Harrogate. This omission was rectified in 1902 with the opening of a south to west curve at Wetherby. A new station at Wetherby, located to the south of the new junction, opened on 1 July 1902, with the original station on the line towards Church Fenton being closed. However, freight traffic continued to be handled at the original site and on 25 September 1963, Class B1 No 61224 is seen at the old station working the pick-up freight from Church Fenton. Following closure in 1902 the original station, designed by G.T. Andrews (who also designed the impressive goods shed visible in this views), was used for residential accommodation; however, it became derelict after the closure of the lines through Wetherby and was demolished in the 1970s. Following the withdrawal of passenger traffic over the lines from Cross Gates and Church Fenton to Harrogate in 1964, the section from Church Fenton to Wetherby remained open for freight traffic until 4 April 1966. The goods shed – now known as 'The Engine Shed' – is still extant, having been converted into an entertainment centre. *Gavin Morrison*

Opposite below: **The station** at Thorp Arch, viewed looking towards the south, is seen here on 12 March 1966; by this date the station had closed completely – passenger services were withdrawn on 6 January 1964 and freight facilities followed on 10 August the same year – but the line remained open for freight traffic – just – to and from Wetherby. Less than a month later, on 4 April, freight facilities were withdrawn from Wetherby and the line west of Tadcaster was closed completely; the rump of the line, from Church Fenton to Tadcaster, succumbed on 30 November 1966. The lines from Church Fenton to Wetherby and from Crossgates to Harrogate (except the section over Crimple Viaduct into Harrogate itself, served by a spur from Pannal), were slated for closure in the Beeching Report of March 1963; there were no objections received for the Tadcaster to Wetherby section and so the inquiry into the closure inevitably gave consent. Fortunately, the station building and goods shed at Thorp Arch, which were both designed by G.T. Andrews, survive, with the station now converted into a private house; both buildings were listed Grade II in 1988. *Neville Stead Collection/Transport Treasury*

THE NORTH EASTERN LINES FROM LEEDS • 151

Above: **Designed by** G.T. Andrews, the station at Tadcaster – seen here from the south on a cold 9 February 1969, just over two years after complete closure (on 30 November 1966 following the withdrawal of freight facilities) and with the track having been lifted – opened on 10 August 1847 with the line from Church Fenton to Harrogate. The station was provided with two platforms and a timber-built trainshed roof. The station would have become a junction had the proposed line from Copmanthorpe to Crossgates being completed in the late 1840s; however, the problems with raising finance in the post-Railway Mania era led to this project failing; however, the line's bridge across the River Wharfe in Tadcaster was completed and the 'junction' created in the form of long siding across the river to serve the Tower Brewery in 1883. Following the closure of the line, the station site was acquired by the local council and, despite opposition, permission to demolish and redevelop the site was given in 1971. The work was completed and there is now little trace of main line through the town; however, the route of the proposed line to Copmanthorpe plus the viaduct over the River Wharfe are still extant and now form part of the Viaduct Walk. *John Meredith/Online Transport Archive*

Opposite above: **On 26 March** 1996, Class 158 158771 has just departed from Garforth station with a westbound service towards Leeds. Garforth was one of the original stations constructed for the opening of the Leeds & Selby Railway in 1834 but the station that exists today is the product of work undertaken in the early 1870s to the design of Thomas Prosser. Garforth was, until 1924, the junction for the private Aberford Railway; to the east of the station was the junction for the line to Castleford via Ledston. Passenger services to Castleford ceased on 22 January 1951 with the section south from Garforth to Allerton Main Colliery closing completely on 14 July 1969. *John Meredith/Online Transport Archive*

Opposite below: **On 21 September** 2019, Class 170 No 170455, still in its ScotRail livery, is seen at Horsforth with a service from Harrogate to Leeds. As with so many stations, the facilities at Horsforth were much reduced during the 1960s and, in 1969, the station became unstaffed. Following this, the station's original buildings were demolished. However, as patronage increased during the 1990s, the decision was taken to enhance the station and during 2002 and 2003 new buildings were constructed on both platforms. These included a new ticket office, opened on 16 July 2003, on the Leeds-bound platforms (and visible on the right of this view), and an enlarged car park on the Harrogate-bound side. In 2012 a turn back facility was installed at Horsforth in order to assist in the future should the timetable be improved. The Class 170s units – of which 139 two- or three-car sets were constructed at Derby Litchurch Lane between 1998 and 2005 – have seen operation with a number of TOCs. ScotRail was originally the largest single user with fifty-five sets, but this number has been reduced; between March and January 2019, sixteen sets – Nos 170453-61/72-78 – were transferred to Arriva Rail North for operation on lines like the Harrogate loop. The demise of Arriva Rail North saw the sets transferred to Northern Trains on 1 March 2020 following the transfer of the franchise to the government's operator of last resort. *Gavin Morrison*

THE NORTH EASTERN LINES FROM LEEDS • 153

LEEDS SOUTH AND EAST

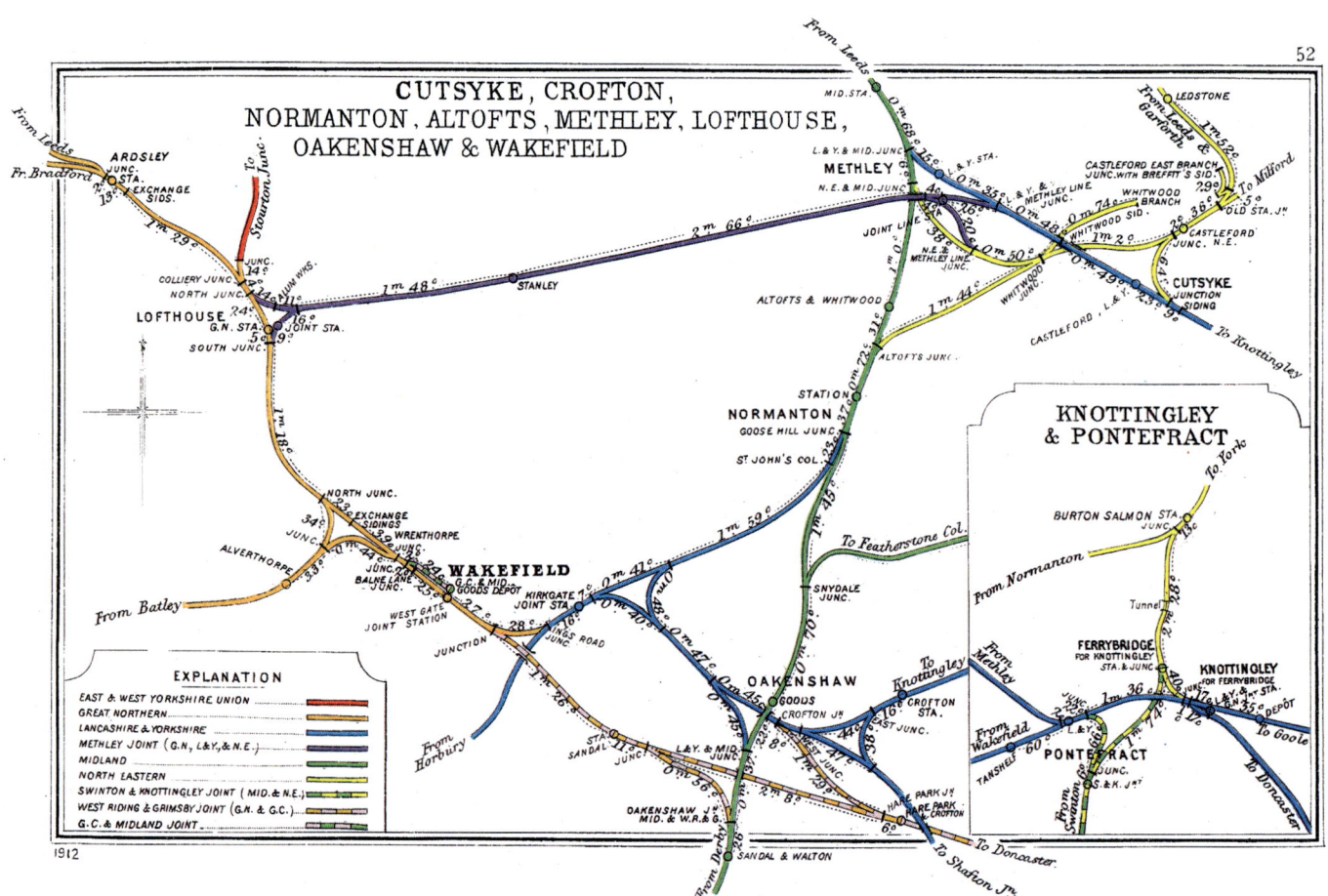

Above: **The Railway** Clearing House map for the Lofthouse and Wakefield district dated 1912. The significant number of industrial sites – in particular collieries – that were a feature of the area are, unfortunately, not shown but were fundamental in the creation of much of the freight traffic on the local railway network.

Opposite above: **The Railway** Clearing House map for the Rothwell and Stourton district in 1912; although Rothwell is shown as a station, by the date the short-lived passenger service to Rothwell had already ceased, the victim of competition from electric trams.

Opposite below: **Recorded heading** west tender first through Woodlesford in 1960 is Class 4F 0-6-0 No 44044. Dating to 1 July 1840, Woodlesford was one of the intermediate stations that opened with the North Midland Railway line from Normanton. When recorded here, the station was still staffed; it was reduced to an unmanned halt in 1970. The station building, situated on the northern platform was demolished the following year. It was not until 2010 that a footbridge was provided at the station; until then passengers wishing to change platforms had to make use of the foot crossing. *Neville Stead Collection/Transport Treasury*

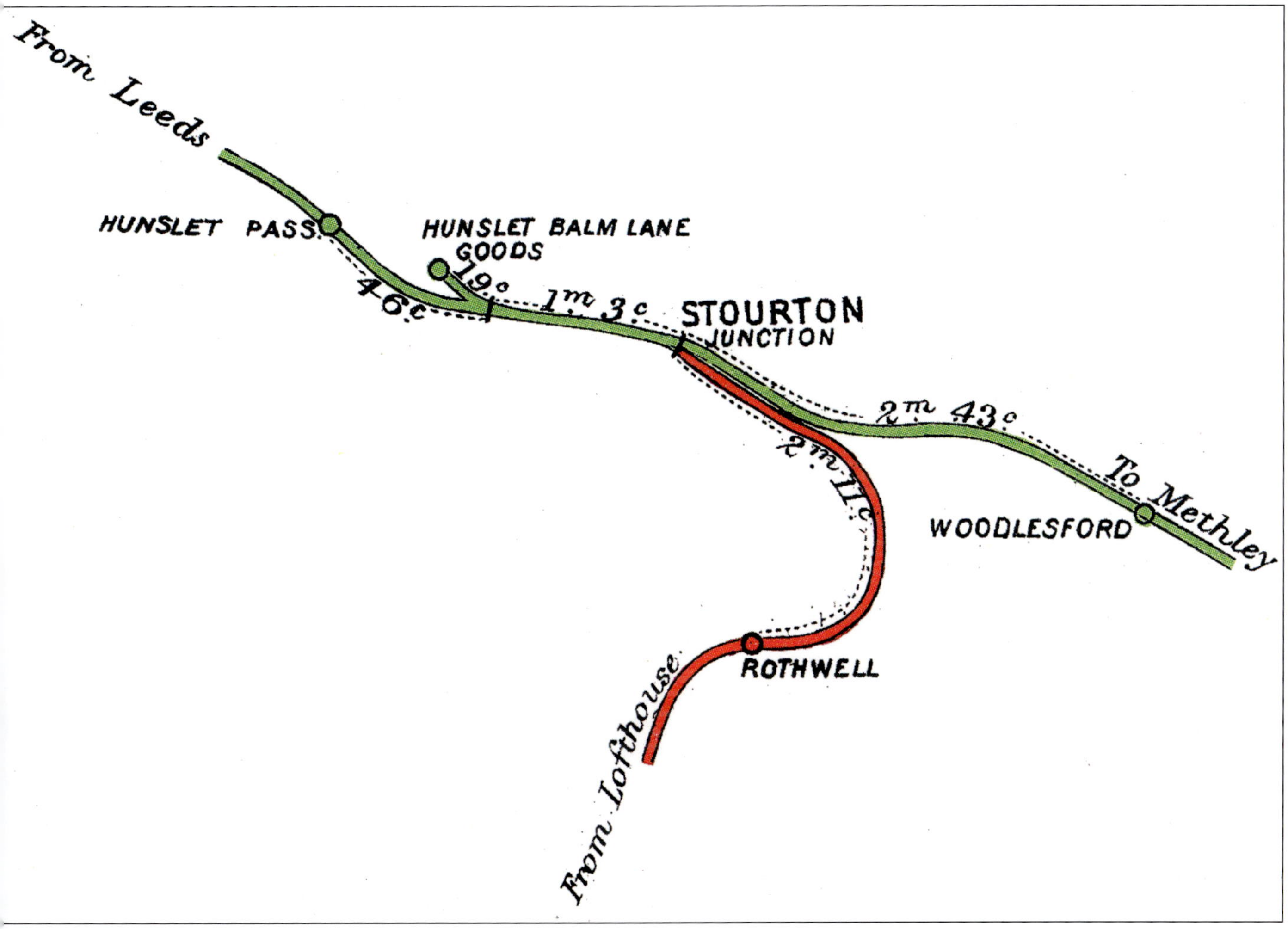

LEEDS SOUTH AND EAST • 155

From Leeds

HUNSLET PASS.
HUNSLET BALM LANE GOODS
46c
19c 1m 3c
STOURTON JUNCTION
2m 43c
2m 11c
WOODLESFORD
To Methley
ROTHWELL
From Lofthouse

On 15 April 1963, Class A4 No 60026 *Miles Beevor* passes through Ardsley station with the Up 'White Rose'. By this date, the 'A4s' were coming towards the end of their life on the East Coast main line with the introduction of the 'Deltic' class diesel-electrics. Indeed, this view of the operation of the 'White Rose' was taken just two months before the service was dieselised; the last steam-hauled 'White Rose' operated on 15 June 1963. Of the thirty-four locomotives that had passed into BR ownership in January 1948, fifteen had been withdrawn by the end of 1963. A number of the class – including No 60026 (which was transferred to St Margaret's, Edinburgh, in October 1963) – were reallocated to Scottish Region for use on the Glasgow to Aberdeen via Stirling and Perth service; this 'Indian Summer' gave the surviving locomotives a new life but was destined to be relatively short-lived, with all being taken out of service by the end of 1966. No 60026 was finally withdrawn in December 1965. *Mike Mitchell/Transport Treasury*

Approaching Ardsley station on 21 August 1968 with the 5pm Saturdays Only service from Bradford Exchange to King's Cross is 2-6-4T No 42116. The lines heading off to the west are the route towards Bradford via Morley and Dudley Hill. Passenger services were withdrawn from the Bradford line on 4 July 1966. The junction was abolished on 5 May 1969 when freight traffic was finally withdrawn from the section to Morley Top. *Mike Mitchell/Transport Treasury*

Above: **The East** & West Yorkshire Union and the South Leeds Junction railways, promoted in the late nineteenth century to help to transport the coal from the collieries south-east of Leeds, briefly saw a passenger service between Stourton and Robin Hood during 1904. However, by that date, the development of the local tramway network meant that the new passenger service was loss-making from the start and services were withdrawn after only nine months. On 21 September 1958, the RCTS organised the 'South Yorkshire No 4' rail tour which ran from Sheffield Victoria and covered a number of lines in the area between Leeds, Sheffield and Doncaster. These included the line from Lofthouse to Rothwell and the train, headed by two ex-LNER 0-6-0s – Class J6 Nos 64222 and 64268 – is pictured awaiting its return from Rothwell southwards. *John McCann/Online Transport Archive*

Opposite above: **Although Kippax** station had lost its passenger services some seven years earlier, the station was still largely intact when, on 21 September 1958, the RCTS 'South Yorkshire No 4' rail tour traversed the Castleford to Garforth line. The train, which was hauled by two Class J6 0-6-0s (Nos 64222 and 64268), is seen heading northbound towards Garforth in the early afternoon; the train was booked to be at Garforth between 2.23pm and 2.33pm before returning southwards. *Neville Stead Collection/Transport Treasury*

Opposite below: **Having run** round their train at Garforth, the two Class J6s then headed back south from Garforth towards Castleford tender first. The train is pictured here at Ledston during its return journey. Dominating the scene is Allerton Bywater Colliery; this coal mine was in operation from 1875 until its final closure in 1992. The southern section of the Garforth to Castleford route, from Ledston southwards, survived until final closure on 6 June 1998, its demise resulting from the closure of the local mines which had provided its only traffic in later years. *Neville Stead Collection/Transport Treasury*

LEEDS SOUTH AND EAST • 159

Above: **On 7 August** 1960, Class 5 No 45187 is seen passing Wards Siding signalbox with the 4.40pm service from Leeds City to Manchester. *Mike Mitchell/Transport Treasury*

Opposite above: **Recorded at** Robin Hood in 1961 with a coal train is Class J6 0-6-0 No 64226. More than five decades after the withdrawal of passenger services which operated north through Rothwell to Stourton and Leeds, the station is still in reasonable condition. Freight traffic continued to use the section from Lofthouse North Junction through Robin Hood to Rothwell until 3 October 1966. The section north from Rothwell to Stourton also closed officially on that date but in reality had been out of use since late 1962. Robin Hood was the junction for the 1¾-mile branch to Royds Green Lower. This line had been sanctioned under the Light Railways Act of 1896 – only the second to be approved (on 14 December 1897) – and opened by the end of the following year. The freight only branch was finally closed on 9 December 1963. Part of the trackbed of the Lofthouse to Rothwell line now forms part of the Rothwell Greenway bridleway. *Neville Stead Collection/Transport Treasury*

Opposite below: **The GNR** line from Adwalton Junction to Batley was opened in two stages during 1863 and 1864. The three-mile route had two intermediate stations – Howden Clough (opened on 1 November 1866) and Upper Batley – and on 8 August 1964, Stanier-designed 2-6-4T No 42650 is pictured from the north with the 12 noon Saturdays Only service from Lowestoft to Bradford Exchange at Howden Clough. Howden Clough station had closed on 1 December 1952 (Upper Batley had closed on 4 February 1952) although the line remained in use for long distance passenger traffic until 7 September 1964; it closed completely on 15 February 1965. *Mike Mitchell/Transport Treasury*

Above: **One of** the factors that prevented the LNWR from quadrupling the existing line from Huddersfield to Leeds was the one mile 1,609 yard long Morley Tunnel, situated immediately to the south of Morley station. Pictured emerging from the tunnel on 10 July 1965 are Fairburn-designed 2-6-4T No 42699 and Stanier Class 5 4-6-0 No 45382 double-heading the afternoon newspaper vans from York to Manchester. *Mike Mitchell/Transport Treasury*

Opposite above: **The station** at Tingley, on the ex-GNR route from Ardsley via Morley Top to Laisterdyke only opened in May 1859, nearly three years after the route itself. The station was closed to timetabled services on 1 February 1954 but was retained for excursion traffic thereafter. However, when recorded here in 1966 towards the end of the line's use in passenger traffic as 2-6-4 No 42116 heads westbound, the station appears derelict with broken windows and an air of terminal decline. *Neville Stead Collection/Transport Treasury*

Opposite below: **On 6 August** 1966, Class 5 4-6-0 No 45191 is pictured approaching Ardsley Tunnel with the 3.5pm serviced from Bradford Exchange to King's Cross. The train is passing under the bridge that once carried the line from Beeston Junction to Tingley. The GN route from Beeston Junction was opened in 1890 with a circular service that ran from Leeds Central via Tingley, Batley and Dewsbury being introduced on 1 November that year, three months after the line's opening; this service operated until 1938. Passenger services from Beeston to Batley via Tingley were withdrawn on 29 October 1951 with the Beeston Junction to Tingley section closing completely on 6 July 1953. *Mike Mitchell/Transport Treasury*

LEEDS SOUTH AND EAST • 163

Seen approaching Morley station from the north on 26 August 1967 is Class 5 No 44940 heading south with Saturdays Only relief service from Darlington to Manchester. The ex-LNWR main line from Leeds has, at this point, been climbing consistently with the final ascent towards Morley being between 1 in 120 and 1 in 138. The summit on this section is in the middle to Morley Tunnel; thereafter the line descends to Mirfield. From there, a further ascent takes the line through Huddersfield and on to Marsden and Standedge Tunnel. *Mike Mitchell/Transport Treasury*

With Lofthouse colliery forming the backdrop, 'Peak' No D45020 heads south on 1 September 1975 with the 17.38 service from Leeds to Bristol. Two years prior to the date of the photograph – on 21 March 1973 – Lofthouse colliery was to be the scene of a tragedy when seven miners were drowned when a new coalface being dug breached the flooded shaft of a disused nineteenth-century working. The colliery, which was rail served, was finally to close in 1981. Classes 45 and 46 were the regular locomotives used on services on the north east/south-west corridor following the elimination of steam until they were displaced by the introduction of HSTs. *Gavin Morrison*

ENGINE SHEDS

Pictured outside Ardsley shed on 5 October 1952 is Class C14 No 67444. The first shed at Ardsley had been opened by the Bradford, Wakefield & Leeds Railway in 1860. This small two-road shed was replaced by a larger eight-road shed by the GNR in 1892. The eastern half of the shed was reroofed after the Second World War, with work on the remainder – very necessary judging by the condition evinced in this view – in 1955. The shed was closed a decade later and subsequently demolished. *Tony Wickens/Online Transport Archive*

Above: **Pictured outside** Copley Hill shed on 5 October 1952 is Class N1 0-6-2T No 69450. Within the triangle formed at Wortley, on the GNR's approach to the city, the LYR and GNR established a joint engine shed in 1857; in 1900 this was replaced by two separate facilities, one serving the LYR to the north (which replaced the original structure) and one to the south for the GNR. Whilst the former was closed on 1 July 1928 and remained in a derelict condition for a period thereafter prior to demolition, the ex-GNR shed, reroofed by BR shortly after nationalisation, was to remain operational until 7 September 1964. It was subsequently demolished. Ivatt's 0-6-2T design was introduced in 1907 and a total of fifty-six were constructed between then and 1912; most were fitted from new with condensing equipment for operation on suburban services to and from London Moorgate station. Four, however, were despatched new to the West Riding in 1912 and other members of the class were also transferred to the area as they were supplanted by newer locomotives. When seen here, No 69450, fitted with condensing apparatus, was a relatively recent arrival, having been reallocated from Hornsey to Copley Hill in June 1951. It was one of seven of the type to survive into 1959. All seven were withdrawn during March and April that year. The 'N1' class was a feature in the operation of many of the ex-GNR routes in the Bradford and Leeds area until their withdrawal. *Tony Wickens/Online Transport Archive*

Opposite above: **The LNWR** opened a twelve-road engine shed at Farnley Junction, situated between the main line towards Huddersfield and the freight-only branch towards Farnley itself, in 1882. The shed was reroofed by the LMS in 1932. Pictured alongside the main shed building on 13 May 1956 is Class 5 4-6-0 No 45075. No 45075 was allocated to Farnley Junction through most of the BR era; transferred to Holbeck in September 1964, the locomotive was finally withdrawn from Normanton in September 1967, following a final reallocation two months earlier. The shed was closed by BR during November 1966 and subsequently demolished. *Eric Sawford/Transport Treasury*

Opposite below: **The MR's** Holbeck shed originally opened on 9 May 1868 and was a double roundhouse structure with two slated triple pitched roofs. However, when recorded here, the shed's great days had gone; it was officially closed by BR on 2 October 1967 and the building demolished soon thereafter. When recorded here – on 31 October 1969 – the site was still in use for the stabling of diesel locomotives. The depot's repair shop was converted to act as a maintenance shed for diesel locomotives and the site remains in use for rolling stock maintenance. *Geoffrey Tribe/Online Transport Archive*

ENGINE SHEDS • 167

Above: **An atmospheric** shot, taken in April 1967 towards the end of the shed's life, sees 2-6-4T No 42689 alongside Class 5 No 44854 amidst the encroaching decay of Holbeck shed. The tank was one the batch designed by Charles Fairburn and was completed at Derby in August 1945; when recorded here, the locomotive was a recent arrival, having been transferred from Royston shed earlier in the month. Its sojourn at Holbeck was destined to be relatively short; a final transfer saw No 42689 move to Low Moor in September 1967, from where it was withdrawn the following month. The 'Black Five' was a long-term resident at Holbeck, having been based there from before the creation of British Railways in January 1948. The closure of the shed saw it move to Normanton in July 1967, from where it was withdrawn three months later. Charles Fairburn, who was the Chief Mechanical Engineer of the LMS following the retirement of William Stanier in 1944 (having held the role in an acting capacity from 1942 when Stanier was seconded to war work), was a native of Bradford and educated at Bradford Grammar School. *John Worley/Online Transport Archive*

Opposite above: **The first** GNR engine shed in Bradford was opened in 1854 in conjunction with the Leeds, Bradford & Halifax Joint Railway. This six-road structure was situated slightly to the east of Adolphus Street station and closed on 11 August 1867 when it was replaced by a new four-road shed situated to the south, on the north side of the new line from Laisterdyke into Bradford Exchange. This itself was to be replaced by a new shed in March 1883, although both it and the original structure of 1854 were to survive in alternative uses until demolition in 1977. The new shed – known as Bowling Junction and latterly as Hammerton Street – was to the south of Hammerton Street Junction. The shed's facilities included a coaling plant – as illustrated in this view of Class N1 No 69443 outside the shed in 1957 – a turntable and repair shop. In the early 1950s, BR decided on the introduction of DMUs; amongst the areas selected for the experimental use of this type of traction was the West Riding and Hammerton Street was selected to provide accommodation. This resulted in the partial reroofing of the depot in 1955 although part of the structure was to remain without a roof through until the building's final closure and eventual demolition in 1991. The site of the shed is now occupied by a bus depot. *Neville Stead Collection/ Transport Treasury*

Opposite below: **Following the** closure of Hammerton Street in 1984, the building remained empty for a couple of years before it was used by the West Yorkshire Transport Museum project for the storage of vehicles acquired for use in the proposed development at Low Moor and the associated Spen Valley line. Amongst the eclectic range of potential exhibits stored for a period at Hammerton Street were ex-Budapest four-wheel tram No 2576 and one of the trio of ex- Central Electricity Generating Board electric locomotives purchased from Spondon Power Station in Derbyshire. The latter was also powered by batteries and so could be used for shunting at Hammerton Street where no overhead was provided. One of the Spondon locomotives, No 3, which had suffered damage prior to purchase, was scrapped but the remaining two passed for further preservation following the failure of the museum project. The tram and locomotive are seen here on 2 November 1986. *Author*

ENGINE SHEDS • 169

Above: **Stored in** Hammerton Street on the same date were one of the museum's two ex-Blackpool trailer cars, No 690; the surviving three-car Class 506 (Manchester-Glossop/Hadfield) EMU; and the two-car battery electric unit (constructed for the Ballater line but used for much of its life on departmental work). Following the demise of the museum project, No 690 was scrapped as was the Class 506 in 1995 (having latterly been based at Butterley) although the battery unit was more fortunate. Having had its asbestos removed – at considerable expense – whilst in the ownership of the museum, it is now based once again on Deeside in Scotland where a short stretch of the Ballater branch has been preserved. *Author*

Opposite above: **The GNR** constructed a two-road engine shed to the north of Ingrow station. This opened in April 1884 and, although closed by the LNER in 1936, remained extant, latterly in private use, until finally final demolished in 1989. The shed is pictured here from the south in 1966. The section of line from Keighley (GN Junction) to Ingrow was finally closed on 28 June 1965 and in the foreground can be seen piled railway sleepers, evidence of recent track lifting. *Neville Stead Collection/Transport Treasury*

Opposite below: **In April** 1967, careworn Fairburn 2-6-4T No 42055 is pictured shunting in the yard in front of the shed at Low Moor. The original LYR shed at Low Moor opened in 1866; this was located adjacent to the station and to the west of the lines heading towards Exchange station. The shed was rebuilt and enlarged in the late 1880s; as rebuilt, it comprised twelve covered roads along with a turntable and combined coaling stage and water tower. Shortly after nationalisation, the LMR reroofed the eastern six tracks but left the other six open to the elements; the building remained in this semi-roofed condition until its final closure on 2 October 1967. New in October 1950, No 42055 was, by April 1967, approaching the end of its life; it was withdrawn at the end of June 1967. *John Worley/Online Transport Archive*

ENGINE SHEDS • 171

Above: **With Manningham** shed in the background, 'Peak' No D106 (later No 45106) is pictured passing through the closed Manningham station with a service towards Forster Square on 31 May 1966. Manningham shed comprised two basic elements. The first, a stone-built roundhouse (as shown here), was completed in 1872; this was supplemented by a four-road shed, constructed in wood, located to the south of the original structure. This was opened in 1887 but had been demolished by the LMS prior to nationalisation. Notice on the extreme right, in the distance, a stabled DMU. Manningham shed was finally closed on 30n April 1967, two years after the closure of the station (on 22 March 1965). *Gavin Morrison*

Opposite above: **Although there** had been locomotive servicing facilities at Mirfield from the late 1840s, it was not until 1885 that the LYR constructed its engine shed on the north side of the main line. The eight-road shed was reroofed by the LMS and it survived in operation until final closure by BR in April 1967. The shed remained in commercial use for a number of years until it was finally demolished in 2007 prior to redevelopment of the site for housing. Pictured outside its home shed on 5 October 1952 is Henry Fowler designed 2-6-4T No 42406. This locomotive was completed at Derby on 10 October 1933 and had been reallocated from Low Moor to Mirfield at the end of June 1949. It was based here until a final transfer at the end of June 1963 saw it move to Wakefield for the final two years of its operational life. *Tony Wickens/Online Transport Archive*

Opposite below: **The NER** constructed a new shed at Neville Hill, to the east of Leeds on the line towards Garforth. This shed, which incorporated four turntables, was opened in October 1894 and, in 1904, the NER's original facility serving Leeds – at Holbeck to the west of Leeds Central station – was closed with its remaining allocation transferred to Neville Hill. In 1960, BR reduced the size of the building, reroofed it and reduced the number of turntables to two. Pictured inside the shed on 10 May 1964 are Class Q8 0-8-0 No 63348 and Class A1 4-6-2 No 60134. The Vincent Raven-designed 'Q8', which dated originally to March 1913 when it was completed at Darlington Works, was approaching the end of its life when recorded here; having been reallocated to Neville Hill in December 1959, it was withdrawn a month after this photograph was taken, on 17 June 1964. No 60134 *Foxhunter*, designed by Arthur Peppercorn, was another product of Darlington Works and was new in November 1948. Based in the Bradford and Leeds area for its entire operational life, the locomotive was reallocated to Neville Hill, its last shed, during the summer of 1963. It was withdrawn in October 1965. *Gavin Morrison*

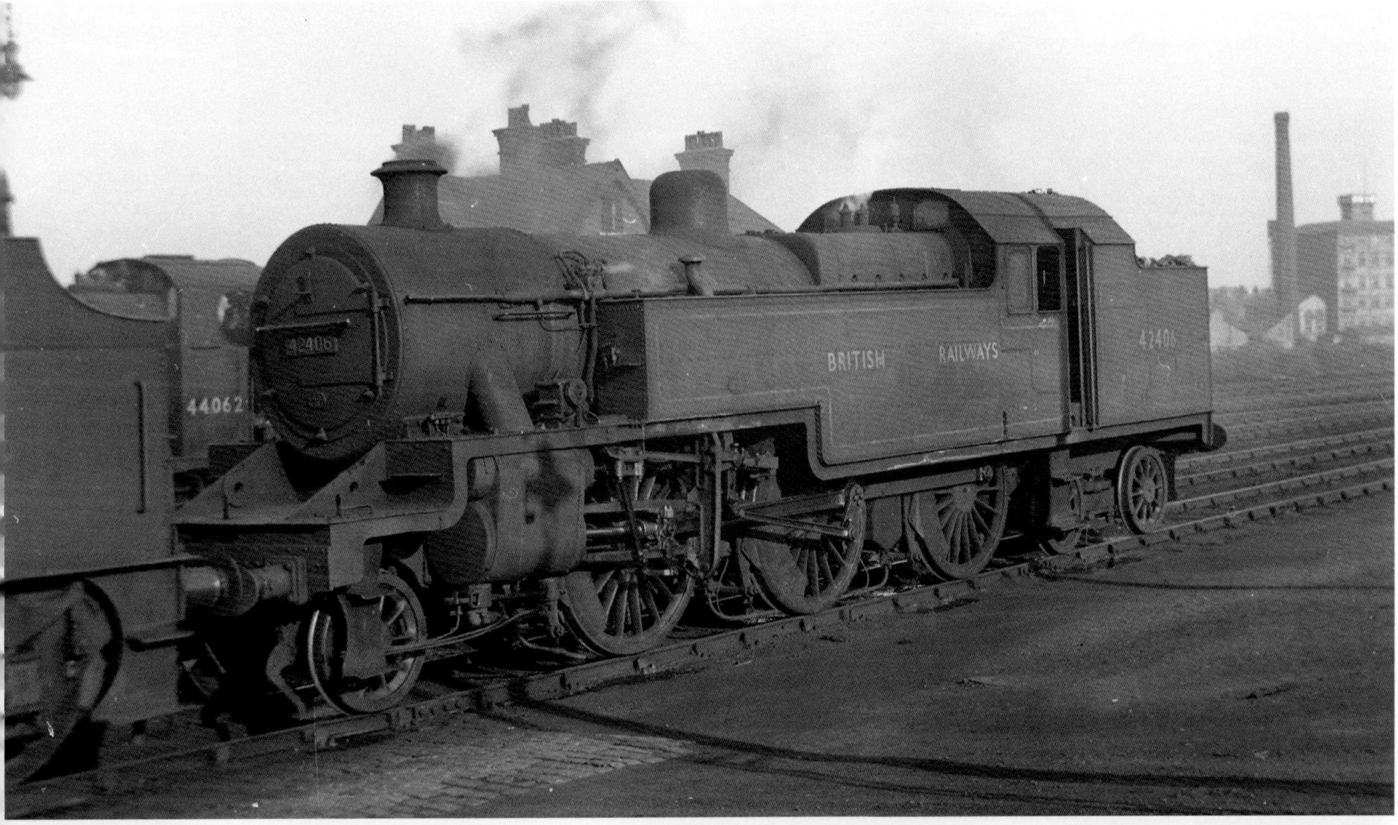
42406
44062
BRITISH RAILWAYS
42406

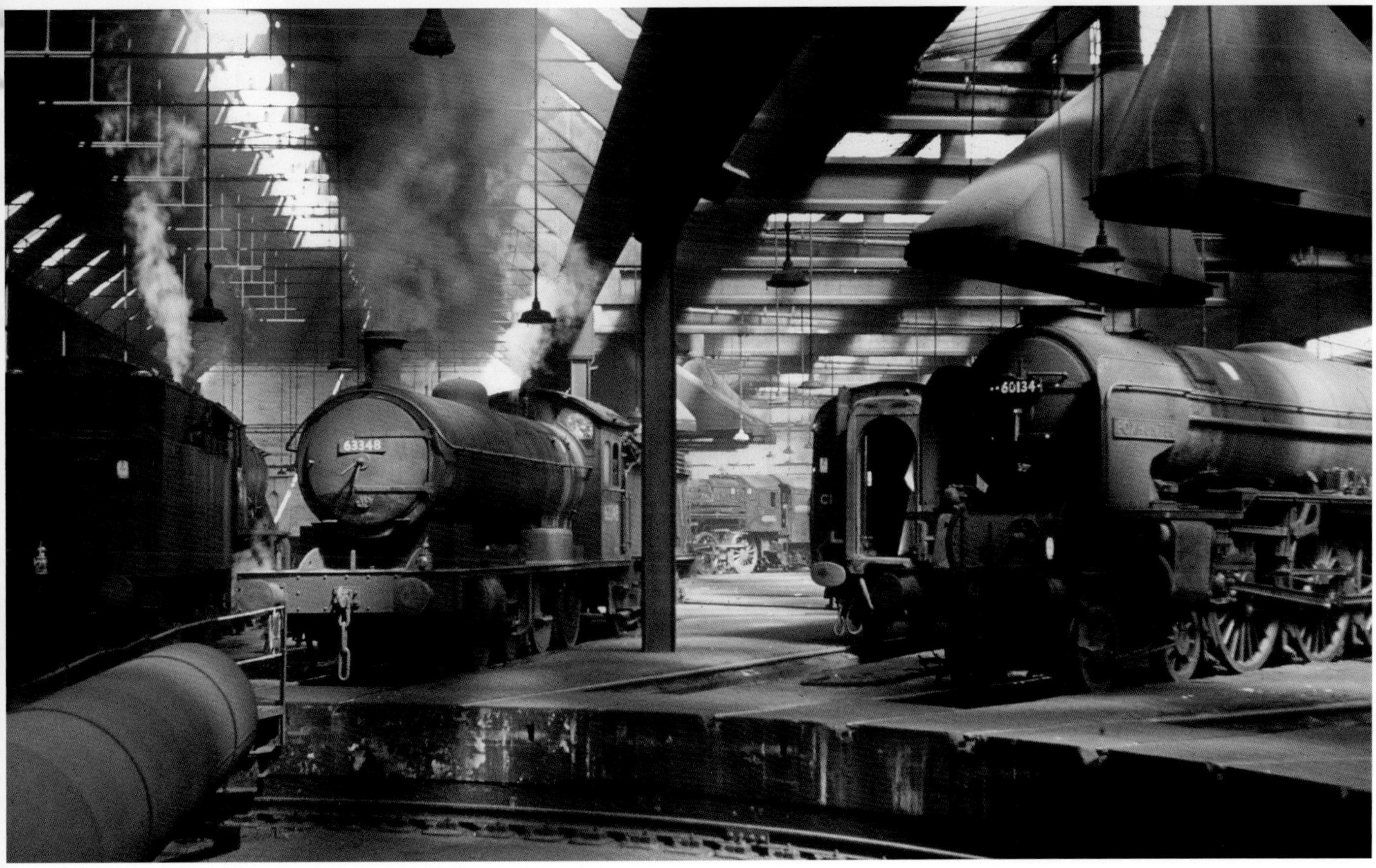
63348
60134

Neville Hill shed was finally closed to steam on 12 June 1966 and the original building was modified to become a simple through shed. A new building, to accommodate DMUs, had already been added – in the late 1950s – and further work was undertaken in 1969 when a completely new set of buildings to cater for diesel locomotives, DMUs and coaching stock was completed. This included an 800ft-long repair shed, which is seen behind HST power car No 43049 on 13 September 1987. HST maintenance commenced at Neville Hill in 1980 and in 1985 no fewer than fifty-five HST power cars were allocated to the depot. Neville Hill remains operational; now owned by Northern Trains, stock from this TOC plus the modern LNER (the franchisee for the East Coast main line at the time of writing) and CrossCountry are stored or maintained at the depot. The last HSTs to be maintained at Neville Hill were those owned by East Midlands Railway until their withdrawal in May 2021. *Gavin Morrison*

On 13 March 1956, Class 3F 0-6-0 43579 is seen outside its then home shed at Stourton. The locomotive was to remain a West-Riding-based locomotive for the remainder of its operational life, being transferred to Wakefield by mid-June 1957, to Low Moor in mid-June 1958 and, finally, back to Wakefield in late October 1959 from where it was withdrawn in late 1960. Stourton shed, which was situated on the west side of the line, was originally opened by the MR in 1893 but had been reroofed by the LMR during 1950. The shed was finally closed in January 1967 and the building demolished. *Eric Sawford/Transport Treasury*

The early years of transport preservation were difficult as vehicles were preserved in an era when secure accommodation was at a premium. The newly-preserved Middleton Railway played host to a number of trams in its early years. The view records no fewer than five trams on the railway in the early 1960s. From the left are two Leeds trams – 'Horsfield' No 202 and railcar No 601 – with, under wraps, ex-Liverpool No 869 (still in its Glasgow livery as No 1055), Sheffield 'Roberts' No 513 and Swansea & Mumbles No 2. Sadly, the broken window on the lower deck of Leeds No 202 evinces the problem that the trams faced in open storage – vandalism. Of the five trams illustrated, fortunately the ex- Liverpool and Sheffield trams survived – the latter spending some time in storage elsewhere in the area (at Cullingworth, Oxenhope and Embsay) before finding a permanent home at Beamish – but the other three were all eventually to be scrapped as a result of damage caused. *F.E.J. Ward/Online Transport Archive*

steady progress was made towards the line's reopening. This was finally achieved on 29 June 1968 when Ivatt-designed 2-6-2T No 41241 and 'USA' 0-6-0T No 72 departed from Keighley station with the first passenger train for more than five years. Since then, the Worth Valley line has become one of the most successful of Britain's preserved railways. Not only does it attract a significant number of passengers annually – thus providing a boost to the local economy – but it has also been hugely successful in films and television work, most notably the much-loved Lionel Jeffries' version of *The Railway Children*, which was filmed on the line during 1970.

The area's third preserved standard gauge line is the Embsay & Bolton Abbey Steam Railway; this project commenced in the late 1960s following the closure of the former MR line from Ilkley to Skipton. A preservation society was formed in 1968 and a base established at Embsay station. A limited shuttle from the station towards Embsay Junction was operated for a number of years. Following the removal of the junction itself, a run-round loop was established at the western end of the now severed section of track. The line's formal reopening took place in 1981 and, since then, the railway has been extended slowly eastwards with Bolton Abbey being reached in 1997, resulting n a line of some four miles in length. The railway has plans for the further extension eastwards to a new station at Addingham, although a return to Ilkley is very unlikely given redevelopment between Addingham and Ilkley and the demolition in 1973 of the bridge immediately to the west of Ilkley station. A further extension is potentially via a reconnected Embsay Junction to return to Skipton station via the existing freight branch to Swinden Quarry. The ex-Ilkley line platforms at Skipton are still extant.

Pictured on the Middleton Railway on 17 September 1967 is ex-NER 0-4-0T No 1310. The NER, which was involved in a number of docks in north-east England, needed to construct a short wheelbase locomotive type for operation in these ports. The Class H (later LNER Class Y7) 0-4-0T was the result. Designed by Thomas Worsdell, a total of twenty-four were constructed between 1888 and 1923; of these nineteen were completed at Gateshead Works with Darlington being responsible for the final five. Only two survived into BR ownership in 1948 – one of which was subsequently preserved – but a number, including No 1310, were sold for industrial use. No 1310 was withdrawn in 1931 and sold to Robert Frazer & Sons; it passed to Pelaw Main Collieries Ltd two years later and thus to the National Coal Board at nationalisation. Acquired for preservation in 1965, it has been based on the Middleton Railway for almost six decades. Although bearing the number 1310, in reality little of No 1310 exists as the frames were those from sister locomotive No 900 fitted with a replacement boiler in 1951 (No 900 had previously been fitted with the boiler from No 1310 and the frames of the latter were subsequently scrapped). *Alexander McBlain/Online Transport Archive*

Above: **On 8 June** 1969 – less than a year after the line opened – ex-BR 'USA' 0-6-0T No 72 heads south from Haworth with a service towards Keighley. This was one of the two locomotives used by the railway to operate the reopening special in June 1968. No 72 – BR No 30072 – was one of fourteen US Army Transportation Corps locomotives acquired by the Southern Railway ion 1947 to replace older tank engines for work on the lines that served Southampton docks. All but one of the fourteen – the exception being No 30061 (which was built by H.K. Porter) – were built by Vulcan Ironworks in the USA. Withdrawal of the class took place between 1962 and 1967 with No 30072 succumbing in July 1967. Three others of the ex-BR locomotives also survive in preservation. No 30072 was rescued from the scrap line at Salisbury by a member of the Keighley & Worth Valley Railway Preservation Society in August 1967; it was based on the line for almost fifty years but has been relocated to the Ribble Steam Railway undergoing a major overhaul following a sale in 2015. *John Worley/Online Transport Archive*

Opposite above: **A main-line** connection brings the potential of through trains from the national network – a facet of railway preservation that the Keighley & Worth Valley has been exploiting since the late 1970s. On 3 November 2012, a special from St Pancras to Oxenhope brought the unusual sight of an East Midland Trains-livered HST to the branch. With power car No 43082, the train is pictured at Oakworth as it makes its way up the valley towards its ultimate destination. Given the line's involvement with the filming of the Lionel Jeffries' version of *The Railway Children* (and the earlier BBC television series based on the same book) the fact that the power car is named *The Railway Children* – after the charity that supports street and vulnerable children found at railway stations is perhaps appropriate. Following withdrawal, No 43082 was secured for preservation by 125 Heritage Ltd and is, at the time of writing, based on the Colne Valley Railway in Essex. *Gavin Morrison*

Originally set up to acquire the Grassington line – the original Yorkshire Dales Railway – in the event of BR deciding to close the route, the preserved Yorkshire Dales Railway (originally the Embsay & Grassington Railway Preservation Society) turned its attention to the closed route from Embsay Junction towards Ilkley when it became apparent that the stone traffic from Rylstone was to be maintained. This view, taken from the footbridge looking towards the west on 1 July 1972, shows the limited steam shuttle that the society was initially able to offer standing in the platform at Embsay station. The society had taken on the lease of the station in 1970 but, with track lifting on the route already well advanced, only managed to secure effectively half-a-mile of track from the junction to a point just east of the station. *John Meredith/Online Transport Archive*

Right: **Taken on** the same occasion, but this time looking to the east towards Bolton Abbey, the closed goods shed at Embsay is visible on the left. Alongside the shed can be clearly seen two of the three preserved trolleybuses – single-deck Liege No 425 and double-deck Bradford No 792 – that were based at Embsay for a number of years. Also clearly visible on the westbound line is the fact that one of the rails has been removed; the railway had acquired a number of examples of metre-gauge rolling stock and had plans to convert a section of running line to metre gauge for operational purposes. *John Meredith/Online Transport Archive*

Below: **One of** the more remarkable achievements in railway preservation in recent years is the restoration of one of the NER's two petrol-electric railcars – No 3170 – on the Embsay & Bolton Abbey Steam Railway. Nos 3170 and 3171 were designed by the NER's then Assistant Chief Mechanical Engineer, Vincent Raven, and were the first passenger carrying vehicles in the world to utilise an internal combustion engine. The two railcars were initially used between Hartlepool and West Hartlepool and between Filey and Scarborough. In later years, they were used on the Cawood branch and, following modification with a larger engine, No 3170 was also used for a period around Harrogate. The two were withdrawn during and 1931. Fortunately, the body of No 3170 was converted into a holiday home, spending more than seventy years near Kirbymoorside before its rescue in 2003. Restoration commenced in 2011 and the restored vehicle was officially launched on 19 October 2018; it is seen here at Embsay on 26 June 2019. *Gavin Morrison*

lawnmowers. Conventional steam locomotives were also constructed from 1888 onwards with almost forty being built, including eight supplied to railways in Ireland, as were steam rollers and other steam-powered road vehicles. Latterly the company concentrated on producing parts for the aircraft industry until its Smithfield Works closed in 1975.

Hudswell Clarke & Co Ltd began life as Hudswell & Clarke in 1860. The company produced steam locomotives – standard gauge, narrow gauge and miniature – between 1860 and 1961; a total of 1,800 were constructed during that period. In the 1920s and 1930s production diversified to include petrol and diesel locomotives; later much of the production was devoted to the construction of engines for use underground in collieries. The company becoming a subsidiary of Hunslet in 1972.

Kitson & Co Ltd was first established in 1837 as Todd, Kitson & Laird and constructed their first steam locomotives the following year; these were two 0-4-2s – *Lion* (now preserved and familiar as the star of the classic Ealing comedy *The Titfield* Thunderbolt) and *Tiger* – for the Liverpool & Manchester Railway. The company went through various partnerships before emerging as Kitson & Co in 1863, becoming a limited liability company in 1886. The company manufactured locomotives both for the domestic and export market but fell into receivership in 1935 by which stage a total of 5,405 locomotives had been constructed. Its patterns and rights were eventually acquired by Robert Stephenson & Hawthorns.

Manning Wardle & Co Ltd grew out of the earlier E.B. Wilson & Co, which had been established at the Railway Foundry, Jack Lane, Hunslet – a street that was also to house Hudswell Clarke and Hunslet – and which had failed in 1858. E.B. Wilson & Co itself was the product of a number of earlier engineering companies. The company primarily concentrated on the production of locomotives for the industrial and contractor market and, by 1900, five years before it became a limited liability company, it had produced more than 1,500 locomotives. However, the company ceased trading in 1927 by which date some 2,000 locomotives had been built. The last locomotive completed – an 0-6-0T (Works No 2047/1926) was built in late 1926 for the cement works at Rugby; following withdrawal, this locomotive was preserved and is now on display at the Kidderminster Railway Museum. The rights to the company's designs were purchased by Kitson & Co; the new owners produced twenty-three examples before it too ceased manufacture.

John Fowler & Co Leeds Ltd is perhaps better known as a manufacturer of agricultural equipment and steam-powered road vehicles but it also produced numerous railway locomotives. It was founded in Bristol in 1850 by John Fowler in association with Albert Fry. In 1856, Fowler moved to London; at this stage he did not manufacture his own products – these were all licensed to other manufacturers (including Kitson & Co) – until 1860 when he established his own factory adjacent to the Kitson's works in Leeds. From 1862 Fowler's Steam Plough Works, in Leathley Lane, produced all of the company's products with the end of the final licensing deal (with Kitson & Co). The first steam locomotive was produced in 1866 – for the Imperial Mexican Railways – and a new factory for the manufacture of locomotives was constructed across the road from the original site. Much of the company's production was for narrow gauge locomotives for use on plantations. Petrol and diesel locomotives were constructed from the early 1920s and the manufacture of steam locomotives ceased at the end of the next decade (although a number of Fowler steam locomotives were built under licence thereafter bearing Fowler Works Numbers). The company became registered as Fowler & Co (Leeds) Ltd in 1886 and merged in 1947 with the Gainsborough-based Marshall Sons & Co Ltd. Production of locomotives ended in 1968 with the business being sold to Andrew Barclay with work at the Steam Plough Works ceasing in 1974.

Robert Hudson Ltd was based at Gildersome, alongside the GNR line, and was established in 1865. In 1875 it patented a successful type of tipper wagon – a product that was produced in large numbers. Although the company produced a number of locomotives itself, most of the locomotives that the company supplied were manufactured under licence by other companies such as Hunslet and Hudswell Clarke in Leeds and the Bristol-based Avonside Engine Co.

Hunslet Engine Co Ltd was established in 1864 and completed its first locomotive – an 0-6-0ST for the contractor Thomas Brassey used on the construction of the MR's extension through Bedfordshire to reach St Pancras station – the following year. One the most famous of the early designs to emerge from the factory was the narrow gauge Quarry Hunslet, built for use in the slate quarries of north-west Wales. The company developed a major export business with locomotives shipped to more than thirty countries by the start of the twentieth century. It became a limited liability company in 1902. The last steam locomotive constructed for use in Britain – prior to the work on replicas such as No 60163 *Tornado* – was one of Hunslet's famous Austerity 0-6-0Ts supplied to Cadeby Colliery in 1964. As other manufacturers disappeared, Hunslet acquired the rights to many of these other companies. Production in later years included a batch of EMUs – the Class 323 – but delivery in 1995 of a batch of locomotives for the construction of the Jubilee Line extension in London marked the end of production at the Hunslet works.

Opened in 1875, Allerton Bywater Colliery was the third mine in the village as new technology permitted deeper shafts and thus the ability to exploit seams that had previously been inaccessible. The mine was originally owned by the Silkstone & Haigh Coal Co but, like the bulk of the industry, passed to the National Coal Board when the industry was Nationalised. Allerton Bywater Colliery was finally closed in March 1992. Pictured at the colliery on 4 October 1952 is Airedale No 3; this locomotive was constructed by the Leeds-based Hunslet Engine Co (Works No 1440) in 1923 for the newly created Airedale Collieries Co Ltd; this company had been created four years earlier by the merger of the Wheldale Coal Co with the then owners of Allerton Bywater Colliery. This locomotive was the first of a number of powerful 0-6-0STs that the company constructed with 15in cylinders that proved popular within the mining industry. In 1963 No 3, by then based at West Riding Colliery (between Whitwood and Altofts), was transferred to Acton Hall Colliery, near Featherstone, before entering preservation in December 1975. The locomotive remains, albeit in an unrestored condition, on the Embsay & Bolton Abbey Steam Railway. *Tony Wickens/Online Transport Archive*

Located just to the south of Woodlesford station, Water Haigh Colliery was a relatively late pit, being first sunk in 1908 when the rights to mine the coal were acquired by the Briggs company. The first coal cut reached the surface on 20 April 1911 and an extensive network of sidings served the site; the coal was moved both by rail and by canal via rail-served wharf on the Aire & Calder Navigation. Nationalised in 1947, the pit continued in production until closure in 1970. Amongst the locomotives used at the colliery was No S101; this was a Hunslet-built 0-6-0T (Works No 1094 of 1912), which is pictured at the colliery on 17 August 1968. *Horace Gamble/ Transport Treasury*

Newmarket (Silkstone) Colliery was situated slightly to the east of Stanley station on the Methley Joint line from Lofthouse to Castleford. The colliery had its origins in the 1830s and by the time it finally closed in 1983 was one of the oldest collieries in the country. Pictured at the colliery on 3 April 1971 is Hunslet 0-6-0ST *Jubilee* (Works No 1726 of 1935). *Horace Gamble/Transport Treasury*

During the late nineteenth century, as Bradford and its industries expanded, so the problem of handling waste generated grew. An initial facility at Frizinghall proved inadequate and so the corporation decided on the purchase of the nearby estate at Esholt and in 1909 parliamentary sanction was gained for the construction of a new sewage works. In order to serve the new work, a standard gauge network (as well as a short section of 2ft 0in gauge line), connected to the MR just to the east of Thackley Tunnel, was constructed. At its peak in 1931, this network extended for some twenty-two miles and employed eleven locomotives, some of which had been converted to operate on oil derived from wool grease – a by-product of the city's staple industry. By the end of the 1950s, the railway had shrunk to only some 6½ miles with two working steam locomotives – *Elizabeth* and *Nellie* – both of which were 0-4-0STs constructed by Hudswell Clarke. On 18 April 1970 one of the duo – *Nellie* (which was new in 1922; Works No 1435) – is seen crossing the bridge over the River Aire on the occasion of a visit organised by the Warwickshire Railway Society. The estate and surviving railway passed from Bradford Corporation to Yorkshire Water in 1975 with dieselisation following the next year. Diesel traction was, however, to be shortlived as the remaining section of the railway closed in 1977. Following withdrawal, *Nellie* was preserved and is now on display in Leeds's Industrial Museum at Armley. Sister locomotive *Elizabeth* (new in 1958, Works No 1888) is also preserved and displayed at Armley. *Horace Gamble/Transport Treasury*

Leeds was home to a significant number of locomotive manufacturers. These included Thomas Green & Sons, Hudswell Clarke & Co Ltd, Kitson & Co Ltd, Manning Wardle & Co Ltd, John Fowler & Co Leeds Limited and Robert Hudson Ltd. The best known was, perhaps, the Hunslet Engine Co Ltd. The company's origins lay in 1864 when it was established at Jack lane, Hunslet by a civil engineer, John Towlerton Leather. For more than 130 years – until 1995 when the Jack Lane site closed following the completion of an order for narrow gauge locomotives for use in the construction of the Jubilee Line extension in London – the works manufactured steam and diesel locomotives for operation both domestically and worldwide. Pictured here under construction in 1955 is an 0-8-0 diesel mechanical locomotive destined for export to the Mufulira Copper Mines in Northern Rhodesia (now Zambia). *Transport Treasury*

APPENDIX

LINE OPENING AND CLOSING DATES

This appendix deals with the opening and closing dates of lines within the Bradford and Leeds area; with the effective cessation of most freight traffic to local yards with the end of, for example, the delivery of domestic coal in the 1980s, a number of extant lines are de facto closed to freight traffic and these are noted as 'line extant' in the tables as are all lines over which freight traffic might be expected at the time of writing. Closure to passenger traffic is much easier to define and so where lines retain a passenger service, these are noted as 'N/A' where appropriate.

ABERFORD RAILWAY

One of the earliest railways in the Bradford and Leeds area, the Aberford Railway was a privately-owned line that linked Garforth, on the Leeds & Selby Railway to Aberford. It was primarily constructed to serve the collieries in the district owned by the Gascoigne family. Richard Oliver Gascoigne, who was also a director of the Leeds & Selby, obtained favourable rates from the larger railway for the shipment of coal produced by his collieries – the collieries were struggling to compete with coal carried by canal to Selby – and the Aberford line was surveyed in 1833 by William Harker and William Walker. As the line traversed private land and was a family venture, no parliamentary powers were required for its construction. The actual date of the line's opening is uncertain, but it was no later than early 1837. Both passengers and freight were carried; the latter was primarily coal mined from the various collieries served. Initially the line was horse or gravity powered but steam traction was introduced in 1870. Following the introduction of steam and the acquisition of a second-hand ex-MR four-wheel coach, the acquisition of which resulted in the end of gravity operation, resulted in improved passenger services. The line continued until the early 1920s; the exhaustion of the many of the seams being mined allied to increased bus competition, where the buses provided a direct link from Aberford to Leeds, resulted in the line's final demise in March 1924.

Section	Opened goods services	Opened passenger services	Closed goods services	Closed passenger services
Aberford to Garforth	Uncertain; between 1835 and 1837 – Leeds & Selby minutes refer to traffic being exchanged before 3 March 1837	Uncertain; suspended for a period from about 1840, when George Hudson's control of the Leeds & Selby resulted in the effective cessation of passenger services on the main line, until 1850 when services on the Leeds & Selby were restored	March 1924	March 1924

BRADFORD, ECCLESHILL & IDLE RAILWAY/IDLE & SHIPLEY RAILWAY

The Bradford, Eccleshill & Idle Railway was authorised on 28 June 1866 to construct the line from Laisterdyke to Idle; backed by the GNR financially, the failure to acquire alternative capital led to it being taken over by the larger company by an Act of 24 July 1871. The same Act saw the GNR take over the Idle & Shipley Railway, which had been authorised on 12 August 1867, to construct the link from Idle to Shipley.

Section	Opened goods services	Opened passenger services	Closed goods services	Closed passenger services
Laisterdyke to Phoenix Works	4 May 1874	15 April 1875	6 August 1979	2 February 1931
Phoenix Works to Idle	4 May 1874	15 April 1875	31 October 1966	2 February 1931
Idle to Shipley	August 1874	15 April 1875	7 October 1968	2 February 1931
Shipley connection to Midland line	1 November 1875	N/A	7 October 1968	N/A
Shipley GN goods branch	August 1874	N/A	7 October 1968	N/A
Cutlers North Junction to Quarry Gap Junction	August 1874	N/A	2 November 1964	N/A

BRADFORD, WAKEFIELD & LEEDS RAILWAY

When the GNR obtained its powers to construct the line from London to York, its proposals for branches to serve Leeds and Sheffield were rejected. Initially, once the GNR main line was open, the railway was able to reach the former via the LYR and MR via Methley to Gelderd Junction, with services commencing on 1 October 1849. This arrangement, which relied upon George Hudson's MR for the section from Methley into Leeds, was unsatisfactory – Hudson was antipathetic towards the GNR – and so an alternative route was sought. This was achieved through the promotion of the Bradford, Wakefield & Leeds Railway, which was authorised on 10 July 1854. The company was empowered to construct a new line – to be operated by the GNR from opening – from Wakefield to Leeds. The GNR could serve the new route via running powers over the LYR. At Leeds the line was linked with the existing GNR-operated Leeds, Bradford & Halifax Junction Railway. Via two Acts – of 23 July 1860 and 17 May 1861 – the company was authorised to construct a new line from Wrenthorpe, north of Wakefield, to connect with the Leeds, Bradford & Halifax Junction Railway line from Adwalton to Batley; the completion of this route provided a third direct link between Bradford and Wakefield. Following an Act of 21 July 1863, the company's name was changed to the West Yorkshire Railway; this was in connection – also authorised on 21 July 1863 – for the construction of the Methley Joint Railway (in conjunction with the LYR and NER). The West Yorkshire Railway and its one-third share of the Methley Joint passed into the ownership of the GNR on 1 January 1865 (confirmed by an Act of 5 July 1865).

Section	Opened goods services	Opened passenger services	Closed goods services	Closed passenger services
Wakefield to Leeds	5 October 1857	3 October 1857	Line extant	N/A
Wrenthorpe South Junction to Wrenthorpe West Junction	6 January 1862	7 April 1862	15 February 1965	7 September 1964
Wrenthorpe West Junction to Roundwood Colliery	6 January 1862	7 April 1862	1 November 1965	7 September 1964
Roundwood Colliery to Ossett (Flushdyke)	6 January 1862	7 April 1862	15 February 1965	7 September 1964
Ossett (Flushdyke) to Ossett	2 April 1864	2 April 1864	15 February 1965	7 September 1964
Ossett to Batley	15 December 1864	15 December 1864	26 March 1956 (this was the section to Shaw Cross colliery; the ex-GNR goods yard was accessed thereafter by a new curve at Batley)	1 July 1909

BRADFORD & THORNTON RAILWAY

Authorised on 24 July 1871, the Bradford & Thornton was taken over by the GNR on 18 July 1872 before services were introduced.

Section	Opened goods services	Opened passenger services	Closed goods services	Closed passenger services
St Dunstan's to Horton Park Junction	November 1876	14 October 1878	26 August 1972	23 May 1955
Horton Park Junction to Clayton	9 July 1877	14 October 1878	28 June 1965	23 May 1955
Clayton to Thornton	April 1878	14 October 1878	28 June 1965	23 May 1955
St Dunstan's (South) Junction to St Dunstan's (East) Junction	November 1876	N/A (rarely used)	26 August 1972	N/A
Horton Park Junction to City Road	November 1876	N/A	26 August 1972	N/A

BRITISH RAIL(WAYS)

The Nationalisation of the railways on 1 January 1948 brought the nation's railways under common ownership for the first time. Under the British Railways Board, part of the British Transport Commission, there were six new regions; three of these were responsible for the operation of the railways in the Bradford and Leeds area. These were, initially, the Eastern Region, which took over effectively the ex-GNR of the LNER lines in the area, the London Midland, which inherited the ex-LMS lines, and the North Eastern, which looked after ex-NER lines of the LNER. Over a period of time, the regional boundaries were altered and, in 1967, the North Eastern Region was abolished and replaced by an enlarged Eastern Region. Following local government reorganisation and the creation of the West Yorkshire Metropolitan County Council, the West Yorkshire Passenger Transport Executive had a role in the funding and development of transport facilities in the area. The Passenger Transport Executive continued after the abolition of the County Council in April 1986 until it was replaced by the West Yorkshire Combined Authority on 1 April 2014. British Rail introduced the brand name 'Inter-City' for long-distance passenger traffic in 1966; operationally, however, trains were still run by the regions until the introduction of Sectorisation in 1982. This resulted in the creation of the new InterCity for long distance trains and Provincial (for other services outside the London and south-east area; Provincial was rebranded as Regional Railways in 1989). Following the passing of the Railways Act 1993, passenger railway services were franchised out; the East Coast main line InterCity service formed one franchise with the other ex-InterCity services in the Bradford and Leeds area passing to the Cross Country franchise with the remaining local services being operated primarily by the successor to Regional Railways in the north. Over the past twenty-five years, there have been a number of different operators for each franchise. At the time of writing, both the East Coast and Northern franchises are being operated by the government's operator of last resort as a result of the failures of the last commercial franchisees whilst the CrossCountry franchise is in the hands of Arriva, a subsidiary of the German state railways (DB). During its existence, BR constructed a handful of lines in the area, primarily spurs to enable traffic to be diverted and permit the closure of sections of other route.

Section	Opened goods services	Opened passenger services	Closed goods services	Closed passenger services
Batley (LNWR) to Batley (GNR) – the Batley south curve	18 April 1966	N/A	1 May 1972	N/A
Heckmondwike to Liversedge (LNWR) – spur from the ex-LYR Spen Valley line to serve oil terminal at Liversedge	1966	N/A	September 1990	N/A

EAST & WEST YORKSHIRE UNION RAILWAY

The late nineteenth century saw a further growth in the railway company ambitions as new entrants sought to break local monopolies. One such railway was the Hull & Barnsley that tried to compete with the NER on coal traffic to Hull. Amongst the lines that the Hull & Barnsley supported was the East & West Yorkshire Union Railway, which planned a route from Drax towards south-east Leeds and the MR and GNR lines that served the area. There were a significant number of collieries in and around Rothwell and the Charlesworth family, which owned a number of them, were also enthusiastic supporters of the proposals. The proposed route, some thirty miles in length, was authorised by an Act of 2 August 1883; however, the company struggled to raise the necessary finance and on 25 June 1886 an Act permitting abandonment was obtained, although a new route – the three miles from Lofthouse to Rothwell – was sanctioned. The line served a number of collieries controlled by the Charlesworth family, with branches often built on land owned by the family (with questions of ownership and rights plaguing the LNER after 1923). A branch from Robin Hood to Royds Green Lower was authorised under the Light Railway Act on 14 December 1897. The East & West Yorkshire Union was operated by the GNR from opening but remained independent until the Grouping in 1923.

Section	Opened goods services	Opened passenger services	Closed goods services	Closed passenger services
Lofthouse North Junction to Robin Hood	20 May 1891	N/A	3 October 1966	N/A
Robin Hood to Rothwell	20 May 1891	4 January 1904	3 October 1966	30 September 1904
Robin Hood to Royds Green Lower	1898	N/A	9 December 1963	N/A

GREAT NORTHERN RAILWAY

The GNR was authorised on 26 June 1846 to construct the main line north from London to York along with the loop from Peterborough via Lincoln to Doncaster. Much of the GNR's significant network was opened through the use of subsidiaries or joint lines but the company itself promoted a number of extensions during the final quarter of the nineteenth century. The GNR became one of the key constituents of the LNER at Grouping in 1923.

Section	Opened goods services	Opened passenger services	Closed goods services	Closed passenger services
Runtlings Lane Junction to Dewsbury Junction	June 1874	9 September 1874	15 February 1965	7 September 1964
Dewsbury Junction to Dewsbury (GN; first station; goods facilities later known as Railway Street)	June 1874	9 September 1874	1 January 1990	15 March 1880
Wrenthorpe North Junction to Wrenthorpe West Junction	March 1875	1 May 1876	1 November 1965	October 1938
Stanningley Junction to new line from Bramley Junction	1877	1 April 1878	1 November 1873	1 November 1873
New line from Bramley Junction to Pudsey Greenside	1877	1 April 1878	6 July 1964	15 June 1964

Section	Opened goods services	Opened passenger services	Closed goods services	Closed passenger services
Queensbury to Holmfield	1 December 1879	1 December 1879	28 May 1956	23 May 1955
Queensbury west curve	1 December 1879	1 December 1879	28 May 1956	23 May 1955
Dewsbury Junction to Batley Carr	15 March 1880	15 March 1880	15 February 1965	7 September 1964
Batley Carr to Batley	12 April 1880	12 April 1880	15 February 1965	7 September 1964
Queensbury to Holmfield	1 December 1879	1 December 1879	28 May 1956	23 May 1955
Thornton to Denholme	1 January 1884	1 January 1884	11 November 1963	23 May 1955
Denholme to Cullingworth	1 April 1884	7 April 1884	11 November 1963	23 May 1955
Cullingworth to Ingrow	1 April 1884	7 April 1884	28 May 1956	23 May 1955
Ingrow to Keighley (GN) Junction	1 April 1884	1 November 1884	28 June 1965	23 May 1955
Keighley (GN) Junction to Keighley (GN) Goods	1 April 1884	N/A	17 July 1961	N/A
Headfield Junction to Dewsbury Junction	October 1887	1 December 1893	Line Partially Extant (closed May 1933; reopened 15 February 1965)	May 1933
Batley to Soothill Wood Colliery	By April 1888	6 July 1953	1 August 1890	29 October 1951
Soothill Wood Colliery to Woodkirk	1 July 1890	6 July 1953	1 August 1890	29 October 1951
Woodkirk to Tingley	1 August 1890	30 June 1964	1 August 1890	29 October 1951
Tingley to Beeston Junction	1 August 1890	6 July 1953	1 August 1890	29 October 1951
Bramley Junction to old line from Stanningley Junction	1 November 1893	1 November 1893	6 July 1964	15 June 1964
Pudsey Greenside to Cutlers South Junction	1 November 1893	1 November 1893	6 July 1964	15 June 1964
Broad Lane Junction to Tyersal Junction	1 December 1893	1 December 1893	1952	October 1938
Dudley Hill Low Moor Junction to Goods Branch Junction	1 December 1894	1 December 1893	1 October 1917	31 August 1914
Goods Branch Junction to Low Moor GN Goods	1 December 1894	N/A	9 May 1933	N/A
Goods Branch Junction to Low Moor GN Junction	1 December 1894	1 December 1893	9 May 1933	31 August 1914
Beeston Junction to Parkside Junction	3 July 1899	N/A	3 July 1967	N/A

Section	Opened goods services	Opened passenger services	Closed goods services	Closed passenger services
Beeston Junction to Parkside (GN) Goods	3 July 1899	N/A	3 January 1966	N/A
Dudley Hill South Curve West Junction to South Curve East Junction	N/A	N/A	N/A	N/A

GUISELEY, YEADON & RAWDON RAILWAY

Authorised on 16 July 1885 to construct the 1¼ mile branch from Guiseley to Yeadon, the company changed its name, following an Act of 5 August 1891, to the Guiseley, Yeadon & Headingley Railway, although the eastward extension was never constructed. Ownership passed to the MR in 1892. Whilst a full branch terminus with platform and station building was constructed, the branch never carried timetabled passenger services, although the occasional excursion did travel along the line.

Section	Opened goods services	Opened passenger services	Closed goods services	Closed passenger services
Yeadon to Rawdon Junction	26 February 1894	N/A	10 August 1964	N/A

HALIFAX & OVENDEN JOINT RAILWAY

Backed financially by the GNR and LYR, the railway was authorised on 30 June 1864; however, progress on the line was slow until it formally became a joint line controlled by the two larger railways on 1 August 1870. It remained a joint line, latterly under the control of the LMS and LNER, until nationalisation on 1 January 1948.

Section	Opened goods services	Opened passenger services	Closed goods services	Closed passenger services
Halifax to Halifax North Bridge	17 August 1874	1 December 1879	1 April 1974	23 May 1955
North Bridge to Holmfield	1 September 1874	1 December 1879	27 June 1960	23 May 1955

HALIFAX HIGH LEVEL & NORTH & SOUTH JUNCTION RAILWAY

Promoted originally as an ambitious scheme to provide a link between the GNR and MR, only three miles were authorised – from Holmfield to Halifax St Paul's – on 7 August 1884. Following an Act of 3 July 1894, the company became jointly controlled by the GNR and LYR; it remained a joint line until 1 January 1948 when it was nationalised.

Section	Opened goods services	Opened passenger services	Closed goods services	Closed passenger services
Homfield to Pellon	1 August 1890	5 September 1890	27 June 1960	1 January 1917
Pellon to Halifax St Paul's	5 September 1890	5 September 1890	27 June 1960	1 January 1917

KEIGHLEY & WORTH VALLEY RAILWAY

Backed by the MR, powers to construct the 4¾-mile branch from Keighley to Oxenhope were obtained by an Act that received the Royal Assent on 30 June 1862. Operated by the MR from opening, the larger company leased the line from 11 August 1876 and assumed ownership from1 July 1881; the transfer was confirmed by an Act that received the Royal Assent on 18 July 1881.

Section	Opened goods services	Opened passenger services	Closed goods services	Closed passenger services
Keighley to Oxenhope	1 July 1867	15 April 1867	18 June 1962	1 January 1962

LANCASHIRE & YORKSHIRE RAILWAY

The LYR was created following an Act of 9 July 1847 by the merger of a number of earlier railways; these were the Ashton, Stalybridge & Liverpool Junction, the Huddersfield & Sheffield Junction, the Liverpool & Bury, the Manchester, Bolton & Bury Canal Navigation & Railway, the Manchester & Leeds, the Wakefield, Pontefract & Goole and the West Riding. The LYR was itself to be merged with the LNWR on 1 January 1922, following an agreement of 25 March 1921. The Railways Act that received the Royal Assent on 19 August 1921 would have imposed a merger in any case as part of the creation of the 'Big Four', but the Act permitted mergers to be completed prior to 1 January 1923; thus, the merger of the LYR and LNWR took effect under Statutory Rules & Orders No 2078. In 1920, shortly before the merger, the LYR network extended over 601 route miles with 291 passenger stations and a total of 1,650 locomotives.

Section	Opened goods services	Opened passenger services	Closed goods services	Closed passenger services
Dewsbury East Junction to Headfield Junction	27 August 1866	1 April 1867	Line Partially Extant; junction with main line severed and and surviving section use to serve the cement terminal	1 December 1930
Headfield Junction to Dewsbury Market Place	27 August 1866	1 April 1867	6 February 1961	1 December 1930
Dewsbury West Junction to Headfield South Junction	27 August 1866	1 April 1867	6 February 1961	1 December 1930
Thornhill to Heckmondwike	10 May 1869	1 June 1869	1986	14 June 1965
Greetland to Stainland & Holywell Green	29 September 1875	1 January 1875	14 September 1959	23 September 1929
Brighouse Anchor Pit Junction to Pickle Bridge	1 March 1881	1 March 1881	4 August 1952	June 1948
Low Moor south curve	22 April 1886	22 April 1886	19 January 1970	1 January 1962

LEEDS & BRADFORD RAILWAY

Proposals for a link between Leeds and Bradford first emerged in the 1830s but it was not until 4 July 1844 that the Leeds & Bradford Railway was authorised. The chairman of the company was the 'Railway King', George Hudson, and the line was leased to the MR from opening. The MR was formally authorised to take over the Leeds & Bradford on 24 July 1851. The 13½-mile route, which was engineered by George Stephenson, was built through the Aire Valley alongside the river and the Leeds-Liverpool Canal to Shipley and thence to Bradford parallel to Bradford Beck and the Bradford Canal; although not the most direct route, this was the easiest in terms of gradients. The most significant engineering work on the line was the 1,496yard long tunnel at Thackley. This was constructed as a result of the river and canal curving sharply around Thackley Hill. Prior to the opening of the line, the extension of the line from Shipley westwards through Keighley and Skipton to Colne was authorised on 30 June 1845 as the Leeds & Bradford (Shipley-Colne Extension) Railway; this line also passed to the MR on 24 July 1851.

Section	Opened goods services	Opened passenger services	Closed goods services	Closed passenger services
Leeds Wellington to Leeds Junction	7 September 1846	1 July 1846	Line Extant	13 June 1966
Leeds Junction to Bradford Market Street	7 September 1846	1 July 1846	Line Extant	N/A
Whitehall Junction to Engine Shed Junction	September 1846	1 July 1846	Line Extant	1901
Shipley (Leeds Junction) to Keighley	16 March 1847	16 March 1847	Line Extant	N/A
Keighley to Skipton	8 September 1847	8 September 1847	Line Extant	N/A
Shipley (Bradford Junction) to Shipley (Bingley Junction)	16 March 1847	16 March 1847	Line Extant	N/A

LEEDS & SELBY RAILWAY

Although there had been tentative plans early in the previous decade, the origins of this line – the first main line to be opened in Yorkshire – lay with the Leeds & Hull Railroad; this company, established in 1824, appointed George Stephenson as engineer and Joseph Locke to undertake the survey. However, a financial crisis in 1825 led to no progress being undertaken and a new scheme – the Leeds & Selby – surveyed by James Walker – was authorised on 1 June 1830. Work commenced on 1 October 1830 with two contractors – Nowell & Sons and Hamer & Pratt – being employed. The line's most significant engineering work was the 700-yard-long Marsh Lane tunnel through Richmond Hill at Leeds. When completed this was the longest railway tunnel in the world. With one single track completed, passenger services were introduced in September 1834; with the completion of the second track three months later, freight traffic commenced. The line became part of the growing national network when, on 29 May 1839, a connection was made at Milford with the first section of the York & North Midland. The latter company, controlled by George Hudson, perceived the Leeds & Selby as a potential competitor for its services into Leeds; in order to thwart this, Hudson engineered a thirty-one-year lease over the Leeds & Selby on 9 November 1840. The lease gave the York & North Midland a right to purchase the railway and was authorised by an Act of 6 April 1841. Although

with the LNWR line at Batley being authorised on 30 June 1862. The company's final authorised extension – that which diverted passenger traffic from the increasingly cramped site at Adolphus Street to the newly enlarged LYR Bradford (suffixed Exchange from 1890) station – was authorised on 18 July 1864. This line's opening permitted passenger services to be withdrawn from Adolphus Street although the impressive station remained in use for freight traffic for more than a century thereafter. The line's acquisition by the GNR was first agreed in April 1863; however, parliament rejected the original Bill in 1864 and it was not until 5 July 1865 that the takeover was authorised. The Leeds, Bradford & Halifax Junction Railway's independent existence ceased on 5 September 1865.

Section	Opened goods services	Opened passenger services	Closed goods services	Closed passenger services
Leeds (Three Signal Bridge Junction) to Holbeck	N/A	1 August 1854	N/A	1 May 1967
Holbeck to Laisterdyke	7 August 1854	1 August 1854	Line Extant	N/A
Laisterdyke to Hammerton Street	7 August 1854	1 August 1854	Line Extant	N/A
Hammerton Street to Bradford Adolphus Street	7 August 1854	1 August 1854	1 May 1972	7 January 1867
Laisterdyke to Bowling Junction	7 August 1854	1 August 1854	1985	9 June 1969
Laisterdyke to Dudley Hill	September 1857 (coal from Adwalton; general goods 1 May 1857)	20 August 1856	December 1979	4 July 1966
Dudley Hill to Birkenshaw & Tong	September 1857 (coal from Adwalton; general goods 1 May 1857)	20 August 1856	16 March 1968	4 July 1966
Birkenshaw & Tong to Gildersome	September 1857 (coal from Adwalton; general goods 1 May 1857)	20 August 1856		4 July 1966
Gildersome to Morley Top	10 October 1857	10 October 1857	16 March 1968	4 July 1966
Morley Top to Ardsley	10 October 1857	10 October 1857	5 May 1969	4 July 1966
Halifax South Parade goods branch	1 September 1856	N/A	c1980	N/A
Adwalton Junction to Upper Batley	19 August 1863	19 August 1983	15 February 1965	7 September 1964
Upper Batley to Batley	1 November 1864	1 November 1864	15 February 1965	7 September 1964
Hammerton Street to Mill Lane Junction	7 January 1867	7 January 1867	Line Extant	N/A

LEEDS, CASTLEFORD & PONTEFRACT JUNCTION RAILWAY

Authorised by an Act of 21 July 1871, the 6¼-mile line from Garforth to Castleford was backed by the NER, with three-quarters of the £160,000 capital required for the line's construction being supplied by the larger railway. Powers to complete the line were transferred to the NER by an Act of 13 July 1876, two years before the route opened.

Section	Opened goods services	Opened passenger services	Closed goods services	Closed passenger services
Garforth to Ledston	8 April 1878	12 August 1878	14 July 1969	22 January 1951
Ledston to Castleford Old Station Junction	8 April 1878	12 August 1878	6 June 1998	22 January 1951

LEEDS CENTRAL

The short – quarter-mile – section from Three Signal Bridge Junction to Leeds Central was originally authorised by an Act of 22 July 1848. Services commenced to a temporary station on 18 September 1848 and the station reached its final form in August 1857.

Section	Opened goods services	Opened passenger services	Closed goods services	Closed passenger services
Leeds Central to Three Signal Bridge Junction	N/A	18 September 1848	N/A	1 May 1967

LEEDS, DEWSBURY & MANCHESTER

Authorised by an Act of 30 June 1845, the Leeds, Dewsbury & Manchester was empowered to construct a line from Leeds to Huddersfield via Dewsbury. The section between Ravensthorpe and Heaton Lodge Junction was the already extant line built by the Manchester & Leeds over which the new railway had running powers. The Act also permitted the construction of the branches to Kirkheaton and from Batley to Birstal. The LNWR was eager to gain access to the West Riding and consequently leased the railway for 999 years on 9 July 1847; whether the Leeds, Dewsbury & Manchester welcomed the attention of the larger railway is a moot point as it held its own – unofficial – opening on 31 July 1848. The LNWR's official opening followed on 18 September 1848. The initial service operated by the LNWR comprised 11 trains per day in each direction between Huddersfield and Leeds; the fast train took 50 minutes between Huddersfield and Leeds with the slowest taking about 75 minutes to complete the journey.

Section	Opened goods services	Opened passenger services	Closed goods services	Closed passenger services
Leeds Three Signal Bridge Junction to Copley Hill Junction	18 September 1848	18 September 1848	18 December 1966	1 March 1882
Copley Hill Junction to North Junction	18 September 1848	18 September 1848	Line Extant	1 March 1882; reopened 18 December 1966
Farnley North Junction to Thornhill Dewsbury Junction	18 September 1848	18 September 1848	Line Extant	N/A
Batley to Birstal	30 September 1852	30 September 1852	18 June 1862	1 January 1917

LONDON & NORTH WESTERN RAILWAY

Known as the 'Premier Line', the LNWR was not only Britain's largest railway company, for a period before the Grouping it was also the largest single company operating in the country. Its origins lay in the merger on 16 July 1846 between three of the earliest mainline railway companies: the Grand Junction, the London & Birmingham and the Manchester & Birmingham. Its influence in the Bradford and Leeds area increased following the construction and opening of the 'Leeds New' line in 1900; this route was designed by the LNWR to increase capacity on its core route from Leeds to Huddersfield; the section south from Heaton Lodge Junction through to Standedge Tunnel had already been quadrupled but quadrupling the existing line via Dewsbury was impractical. The LNWR merged with the LYR on 1 January 1922 before becoming the largest constituent of the LMS on 1 January 1923.

Section	Opened goods services	Opened passenger services	Closed goods services	Closed passenger services
Farnley Junction to Farnley	2 April 1866 (junction altered to east facing 31 May 1885)	October 1997	N/A	N/A
Canal Junction to Farnley North Junction	1 March 1882	1 March 1882	18 December 1966	18 December 1966; reopened from new Gelderd Road Junction on ex-GNR line from Wakefield to Canal Junction 1 May 1967; closed 11 October 1987 with line severed at Canal Junction in connection with electrification of the East Coast main line
Spen Valley Junction to Northorpe	18 September 1899	1 October 1900	11 January 1966	2 August 1965
Northorpe to new junction with ex-LYR route at Heckmondwike	9 July 1900	1 October 1900	11 January 1966	2 August 1965
Junction with ex-LYR route Heckmondwike to Liversedge	9 July 1900	1 October 1900	September 1990	2 August 1965
Liversedge to Farnley Junction	9 July 1900	1 October 1900	11 January 1966	2 August 1965

MANCHESTER & LEEDS

Authorised by an Act on 4 July 1836, the Manchester & Leeds was empowered to construct some 50 miles of line from Manchester via Todmorden and Wakefield to Goose Hill Junction, south of Normanton, where a connection was made with the North Midland Railway to gain access to Leeds. Engineered by George Stephenson (with Thomas Gooch, brother of the Great Western's Daniel, as his assistant), the line opened from Manchester to Littleborough on 3 July 1839 and from Goose Hill Junction to Hebden Bridge on 4 October 1840. The link through the one mile 125 yard Summit Tunnel was completed in March 1841. The Manchester & Leeds merged with a number of other railways to become the Lancashire & Yorkshire on 9 July 1847.

Section	Opened goods services	Opened passenger services	Closed goods services	Closed passenger services
Manchester to Littleborough	3 July 1839	3 July 1839	Line Extant	N/A
Normanton (Goose Hill Junction) to Heaton Lodge Junction	4 October 1840	4 October 1840	Line Extant	N/A
Heaton Lodge Junction to Sowerby Bridge	4 October 1840	4 October 1840	Line Extant	N/A; closed 5 January 1970; reopened XXXX
Sowerby Bridge to Hebden Bridge	4 October 1840	4 October 1840	Line Extant	N/A
Hebden Bridge to Summit Tunnel (east)	31 December 1841	31 December 1841	Line Extant	N/A
Summit Tunnel (east) to Littleborough	1 March 1841	1 March 1841	Line Extant	N/A
North Dean to Halifax Shaw Syke	1 April 1848	1 April 1848	Line Extant	N/A

METHLEY JOINT RAILWAY

Originally promoted by the Bradford, Wakefield & Leeds Railway, the line was authorised by an Act of 21 July 1863. However, the plans had been heavily opposed by both the L&R and NER and, in order to further the GNR's ambitions in the West Riding (in particular for the takeover of both the Bradford, Wakefield & Leeds and the Leeds, Bradford & Halifax Junction railways), a further Act of 23 June 1864 saw ownership of the line from Lofthouse & Outwood to Methley vested in the West Yorkshire Railway (as the Bradford, Wakefield & Leeds had become on 21 July 1863), the LY& and NER in equal thirds. With the West Yorkshire Railway becoming part of the GNR in 1865, the ownership of the West Yorkshire Railway's third passed to the large company. With the LYR passing to the LMS in 1923 and both the GNR and NER to the LNER, joint ownership of the line persisted through until nationalisation in January 1948.

Section	Opened goods services	Opened passenger services	Closed goods services	Closed passenger services
Lofthouse & Outwood to Newmarket (Silkstone)	June 1865	1 May 1869	54 April 1965	2 November 1964
Newmarket (Silkstone) to Methley	June 1865	1 May 1869	23 February 1981	2 November 1964

MIDLAND RAILWAY

Formed by the merger of the Midland Counties, the Birmingham & Derby junction and the North Midland railways on 10 May 1844, the MR expanded its network with the construction of the route via Ilkley to Skipton and, most notably, it developed plans for the construction of a cut-off route that would have resulted in its main line heading south from Shipley via Bradford to connect with its existing Leeds to Sheffield main line at Royston. The new route was authorised via the Midland Railway (West Riding Lines)

Act of 1898 and by additional legislation between then and the outbreak of the First World War. Although the southern section of the route was completed before the start of hostilities in 1914, the extension north from Thornhill via the Spen Valley and Bradford to Shipley was postponed and, despite promises that work would recommence after the war, the through route was never completed. The project was formally abandoned on 18 November 1919 and the land and property that the MR had acquired in the city centre to facilitate construction was sold to the corporation the following year. The desire for a through route has surfaced on several occasions since 1920. The plans for the wholesale redevelopment of the city centre in the early 1960s saw plans revived, although the perception that the railways were in decline perhaps led to no progress at the time, whilst in the early 1980s West Yorkshire Metropolitan County Council employed consultants to examine the future of the local railway network. Amongst one of the options proposed was one that planned a considerable restoration of closed routes as well as a new link in Bradford. This again was not progressed. At the date of this volume's compilation, there are new plans for the construction of a high-speed link between Manchester and Leeds via Bradford; whether, after almost two centuries, Bradford will finally achieve a through route, only time will tell!

Section	Opened goods services	Opened passenger services	Closed goods services	Closed passenger services
Apperley Junction to Burley Junction	1 October 1866	1 August 1865	Line Extant	N/A
Menston Junction to Milner Wood Junction	1 October 1866	1 August 1865	5 July 1965	25 February 1957
Shipley Guiseley Junction to Esholt Junction	13 December 1876	4 December 1876	Line Extant	N/A
Skipton to Bolton Abbey	27 August 1888	16 May 1888	5 July 1965	22 March 1965
Bolton Abbey to Embsay Junction	1 October 1888	1 October 1888	5 July 1965	22 March 1965
Embsay Junction to Skipton	1 October 1888	1 October 1888	Line Extant	22 March 1965
Royston Junction to Crigglestone	3 July 1905	1 July 1909	4 May 1968	13 June 1960; closed 1 January 1917; reopened 3 May 1920; closed 1946; reopened 4 January 1960
Crigglestone to Thornhill Junction	10 November 1905	1 July 1909	4 May 1968	13 June 1960; closed 1 January 1917; reopened 3 May 1920; closed 1946; reopened 4 January 1960

NORTH EASTERN RAILWAY

The NER was created through the merger on 31 July 1854 of the York & North Midland Railway, the Leeds Northern Railway and the York, Newcastle & Berwick railways.

Section	Opened goods services	Opened passenger services	Closed goods services	Closed passenger services
Arthington South Junction to Otley	1 October 1866	1 February 1865	5 July 1965	22 March 1965
Arthington West Junction to Arthington North Junction	1 October 1866	1 August 1877	5 July 1965	25 February 1957
Cross Gates to Wetherby East Junction	1 May 1876	1 May 1876	27 April 1964	6 January 1964
Neville Hill West Junction to Hunslet (NE Goods)	2 January 1899	N/A	Line Extant	N/A
Branch off line to Harrogate to Cardigan Road (Goods), Leeds	19 May 1900	N/A	4 September 1972	N/A
Wetherby South Junction to Wetherby West Junction	1902	1902	6 January 1964	6 January 1964

NORTH MIDLAND RAILWAY

The Act authorising the construction of the North Midland Railway from Derby via Rotherham to Leeds received the Royal Assent on 4 July 1836. The route was engineered by George Stephenson, although the actual construction was managed by Frederick Swanwick (one of Stephenson's assistants), and at Normanton it was designed to form a connection with both the Manchester & Leeds and the York & North Midland railways. The route opened from Derby to Rotherham, avoiding Sheffield (the link through from Chesterfield to Sheffield was not opened until 1870), on 11 May 1840 and thence via Normanton to Leeds on 1 July 1840. Following a merger with the Midland Counties and the Birmingham & Derby Junction railways on 10 May 1844, the railway became the Midland Railway. The MR became part of the LMS at Grouping in 1923.

Section	Opened goods services	Opened passenger services	Closed goods services	Closed passenger services
Normanton to Leeds	1 July 1840	1 July 1840	Line Extant	N/A

OTLEY & ILKLEY JOINT RAILWAY

Promoted by the Midland and North Eastern, the Otley & Ilkley Joint was authorised on 11 July 1861. It remained a joint line through until the nationalisation of the railway on 1 January 1948, being vested in the LMS and LNER from Grouping.

Section	Opened goods services	Opened passenger services	Closed goods services	Closed passenger services
Otley to Burley Junction	1 October 1866	1 August 1865	5 July 1965	22 May 1965
Burley Junction to Ilkley	1 October 1866	1 August 1865	Line Extant	N/A

SOUTH LEEDS JUNCTION RAILWAY

Promoted by the East & west Yorkshire Union Railway, this two-mile section provided a link between the existing railway at Rothwell with Stourton on the MR main line into Leeds. The line was authorised on 24 August 1893. It designed to exploit further the coal mines south-east of Leeds and was operated initially by the East & west Yorkshire Union Railway, which formally absorbed the company following an Act of 2 July 1896.

Section	Opened goods services	Opened passenger services	Closed goods services	Closed passenger services
Rothwell to Stourton sidings	6 April 1895	4 January 1904	3 October 1966 (officially; unused from February 1962)	30 September 1904
Connection with MR at Stourton	1 November 1903	4 January 1904	3 October 1966 (officially; unused from February 1962)	30 September 1904

WEST RIDING UNION RAILWAYS

Formed by the merger of the earlier West Yorkshire and the Leeds & West Riding Junction railways, the West Riding Union Railways – the word 'Railways' was used in its Act – was authorised by an Act that received the Royal Assent on 18 August 1846. The Act also stipulated that the company was to be merged with the LYR within three months and this was achieved on 17 November 1846.

Section	Opened goods services	Opened passenger services	Closed goods services	Closed passenger services
Mirfield to Heckmondwike Junction	18 July 1848	18 July 1848	14 June 1965	14 June 1965
Heckmondwike Junction to Low Moor	18 July 1848	18 July 1848	XXXXX	14 June 1965
Bradford Exchange to Low Moor	9 May 1850	9 May 1850	Line Extant	N/A (line cut back from Exchange to new station 15 January 1973)
Halifax to Low Moor	7 August 1850	7 August 1850	Line Extant	N/A
Milner Royd Junction to Dryclough Junction	1 January 1852	1 January 1852	Line Extant	N/A

YORK & NORTH MIDLAND RAILWAY

The York & North Midland received its original Act on 21 June 1836 to construct a line from York to connect with the Leeds & Selby Railway (opened in 1839) and to the North Midland Railway at Normanton (in 1840). Powers to construct a branch from Church Fenton to serve Harrogate via Tadcaster and Wetherby were obtained in 1845. The York & North Midland became one of the constituents of the NER on 31 July 1854.

Section	Opened goods services	Opened passenger services	Closed goods services	Closed passenger services
Church Fenton to Tadcaster	10 August 1847	10 August 1847	30 November 1966	6 January 1964
Tadcaster to Wetherby	10 August 1847	10 August 1847	4 April 1966	6 January 1964
Wetherby to Spofforth	10 August 1847	10 August 1847	6 January 1964	6 January 1964
Spofforth to Crimple	20 July 1848	20 July 1848	6 January 1964	6 January 1964

YORKSHIRE DALES RAILWAY

During the railway age numerous railways started out with bold ambitions only to end up being a relatively insignificant branch. Such a railway was the Yorkshire Dales; originally authorised on 6 august 1897 to construct a line from Embsay Junction to Grassington – a distance of 8¾ miles – the railway had plans, never fulfilled, to reach Darlington. Although operated by the MR from opening, the company retained its independence until Grouping in 1923.

Section	Opened goods services	Opened passenger services	Closed goods services	Closed passenger services
Embsay Junction to Swinden Siding (Rylstone)	29 July 1902	29 July 1902	Line Extant	22 September 1930
Swinden Siding (Rylstone) to Grassington	29 July 1902	29 July 1902	11 August 1969	22 September 1930

BIBLIOGRAPHY

Bairstow, J.M.; *Railways of Keighley*; Dalesman Publishing Co; 1979
Bairstow, Martin; *The Great Northern Railway in the West Riding*; Author; 1999
Baker S.K.; *Rail Atlas Great Britain and Ireland*; 15th edition; OPC; 2020
Baker, S.; *Rail Atlas of Britain 1977*; OPC; 1977
Batty, Stephen R.; *Rail Centres: Leeds/Bradford*; Ian Allan Ltd; 1989
Baughan, Peter E.; The Railways of Wharfedale; David & Charles; 1969
Biddle, Gordon; *Britain's Historic Railway Buildings: A Gazetteer of Structures and Sites*; Ian Allan Publishing; 2011
British Railways Board; *The Reshaping of British Railways*: Her Majesty's Stationery Office; 1963
Chapman, Stephen; *Railway Memories 11: Airedale & Wharfedale*; Bellcode Books; Undated
Clinker, C.R.; *Clinker's Register of Closed Passenger Station and Goods Depots*; Avon Anglia; 1988
Griffiths, Roger, and Smith, Paul; *The Directory of British Engine Sheds and Principal Locomotive Servicing Points: 2 – North Midlands, Northern England and Scotland*; OPC; 2000
Hoole, Ken (Editor); *Tomlinson's North Eastern Railway*; David & Charles; 1987 (original book published 1914)
Hudson, Graham S.; *The Aberford Railway and the History of the Garforth Collieries*; David & Charles; 1971
Hurst, Geoffrey; *Register of Closed Railways*; Milepost Publications; 1992
Joy, David; *A Regional History of the Railways of Great Britain: Volume 8 – South and West Yorkshire*; David & Charles; 1975
Marshall, John; The Lancashire & Yorkshire Railway: Volume 1; David & Charles; 1969
Marshall, John; The Lancashire & Yorkshire Railway: Volume 1; David & Charles; 1970
Marshall, John; The Lancashire & Yorkshire Railway: Volume 3; David & Charles; 1972
Quick, Michael; *Passenger Stations in Great Britain – A Chronology*; fourth edition; Railway & Canal Historical Society; 2009
Waller, Peter; *Celebration of Steam: West Riding*; Ian Allan Publishing; 1994
Waller, Peter; *Rail Atlas: The Beeching Era*; Ian Allan Publishing; 2013
Whitaker, Alan, and Cryer, Bob; *The Queensbury Lines: A Pictorial Centenary Edition*; Dalesman; 1984
Whitaker, Alan, and Myland, Bryan; *Railway Memories 4: Bradford*; Bellcode Books; 1993
Whitaker, Alan; *Bradford Railways Remembered*; Dalesman; 1986
Wild, Jack, and Chapman, Stephen; *Railway Memories 11: Halifax and the Calder Valley*; Bellcode Books; 1998